WEST-E
0435

General Science
Teacher Certification Exam

By: Sharon Wynne, M.S.
Southern Connecticut State University

"And, while there's no reason yet to panic, I think it's only prudent that we make preparations to panic."

MANKOFF

XAMonline, INC.
Boston

To obtain permission(s) to use the material from this work for any purpose including workshops or seminars, please submit a written request to:

XAMonline, Inc.
21 Orient Ave.
Melrose, MA 02176
Toll Free 1-800-509-4128
Email: info@xamonline.com
Web www.xamonline.com
Fax: 1-781-662-9268

Library of Congress Cataloging-in-Publication Data

Wynne, Sharon A.
    General Science 0435 Teacher Certification / Sharon A. Wynne. -2nd ed.
    ISBN 978-1-58197-634-2
    1. General Science 0435   2. Study Guides.   3. WEST
    4. Teachers' Certification & Licensure.   5. Careers

Managing Editor   Dr. Harte Weiner, Ph.D.
Senior Editor   Zakia Hyder

Disclaimer:
The opinions expressed in this publication are the sole works of XAMonline and were created independently from the National Education Association, Educational Testing Service, or any State Department of Education, National Evaluation Systems or other testing affiliates.

Between the time of publication and printing, state specific standards as well as testing formats and website information may change that is not included in part or in whole within this product. Sample test questions are developed by XAMonline and reflect similar content as on real tests; however, they are not former tests. XAMonline assembles content that aligns with state standards but makes no claims nor guarantees teacher candidates a passing score. Numerical scores are determined by testing companies such as NES or ETS and then are compared with individual state standards. Passing scores vary from state to state.

Printed in the United States of America   œ-1

WEST-E: General Science 0435
ISBN: 978-1-58197-634-2

TABLE OF CONTENTS

COMPETENCY 3.0   THE LIFE SCIENCES

Great Study and Testing Tips!

What to study in order to prepare for the subject assessments is the focus of this study guide, but equally important is *how* you study.

You can increase your chances of truly mastering the information by taking some simple, but effective steps.

Study Tips:

1. <u>Some foods aid the learning process</u>. Foods such as milk, nuts, seeds, rice, and oats help your study efforts by releasing natural memory enhancers called CCKs (*cholecystokinin*), which is composed of *tryptopha*n, *choline*, and *phenylalanine*. All of these chemicals enhance the neurotransmitters associated with memory. Before studying, try a light, protein-rich meal of eggs, turkey, or fish. All of these foods release the memory enhancing chemicals. The better the connections, the more you comprehend.

Likewise, before you take a test, stick to a light snack of energy boosting and relaxing foods. A glass of milk, a piece of fruit, or some peanuts all release various memory-boosting chemicals and help you to relax and focus on the subject at hand.

2. <u>Learn to take great notes</u>. A byproduct of our modern culture is that we have grown accustomed to getting our information in short doses (i.e. TV news sound bites or newspaper articles of *USA Today* style)**.** Consequently, we have subconsciously trained ourselves to assimilate information in <u>neat little packages</u>. If your notes are scrawled all over the paper, it fragments the flow of the information. Strive for clarity. Newspapers use a standard format to achieve clarity. Your notes can be much clearer through use of proper formatting. A very effective format is called the "Cornell Method."

> Take a sheet of loose-leaf lined notebook paper and draw a line all the way down the paper, about 1-2" from the left-hand edge.
>
> Draw another line across the width of the paper, about 1-2" up from the bottom. Repeat this process on the reverse side of the page.

Look at the highly effective result. You have ample room for notes, a left hand margin for special emphasis items or inserting supplementary data from the textbook, a large area at the bottom for a brief summary, and a small rectangular space for just about anything you want.

3. <u>Get the concept, then the details</u>. Too often, we focus on the details and do not understand the concept. However, if you only memorize dates, places, or names, you may well miss the whole point of the subject.

A key method to understanding ideas is to put them in your own words. If you are working from a textbook, automatically summarize each paragraph in your mind. If you are outlining text, don't simply copy the author's words.

Rephrase ideas or concepts in your own words. You remember your own thoughts and words much better than someone else's, and subconsciously tend to associate the important details to the core concepts.

4. <u>Ask "Why?"</u> Pull apart written material paragraph by paragraph and don't forget to take note of the captions under the illustrations.

For example: If the heading is "Stream Erosion," flip it around to read "Why do streams erode?" Then answer the question.

If you train your mind to think in a series of questions and answers, not only will you learn more, you will also reduce test anxiety because you are now used to answering questions.

5. <u>Read for reinforcement and future needs</u>. Even if you only have 10 minutes, keep your notes or a book in your hand. Your mind is similar to a computer; you have to input data in order to have it processed. *By reading, you are creating the neural connections for future retrieval.* The more times you read something, the more you reinforce the learning of ideas.

Even if you don't fully understand something on the first pass, *your mind stores much of the material for later recall.*

6. <u>Relax to learn, so go into exile</u>. Our bodies respond to an inner clock called biorhythms. Burning the midnight oil works well for some people, but not everyone.

If possible, set aside a particular place to study that is free of distractions. Turn off the television, cell phone, pager and exile your friends and family during your study period.

If you are bothered by silence, try turning on background music. Light classical music at a low volume has been shown to aid in concentration over other types. Music that evokes pleasant emotions without lyrics are highly suggested. Try just about anything by Mozart. It relaxes you.

7. Use arrows, not highlighters. At best, it's difficult to read a page full of yellow, pink, blue, and green streaks. Try staring at a neon sign for a while and you'll soon see that the horde of colors obscure the message.

A quick note, a brief dash of color, an underline, and an arrow pointing to a particular passage is much clearer than a horde of highlighted words.

8. Budget your study time. Although you shouldn't ignore any of the material, *allocate your available study time in the same ratio that topics may appear on the test.*

Testing Tips:

1. Get smart, play dumb. Don't read anything into the question. Don't make an assumption that the test writer is looking for something else than what is asked. Stick to the question as written and don't read extra things into it.

2. Read the question and all the choices *twice* before answering the question. You may miss something by not carefully reading, and then re-reading both the question and the answers.

If you really don't have a clue about the right answer, leave it blank on the first time through. Go on to the other questions, as they may provide a clue as to how to answer the skipped questions.

If later on, you still can't answer the skipped ones . . . *Guess.* The only penalty for guessing is that you *might* get it wrong. Only one thing is certain; if you don't put anything down, you will get it wrong!

3. Turn the question into a statement. Look at the way the questions are worded. The syntax of the question usually provides a clue. Does it seem more familiar as a statement rather than as a question? Does it sound strange?

By turning a question into a statement, you may be able to spot whether an answer sounds right; this may also trigger memories of material you have read.

4. Look for hidden clues. It's actually very difficult to compose multiple-choice questions without giving away part of the answer in the options presented.

In most multiple-choice questions you can often readily eliminate one or two of the potential answers. This leaves you with only two real possibilities and automatically your odds go to fifty-fifty for very little work.

5. Trust your instincts. For every fact that you have read, you subconsciously retain something of that knowledge. On questions that you aren't really certain about, go with your basic instincts. **Your first impression on how to answer a question is usually correct.**

6. Mark your answers directly on the test booklet. Don't bother trying to fill in the optical scan sheet on the first pass through the test.

Just be very careful not to miss-mark your answers when you eventually transcribe them to the scan sheet.

7. Watch the clock! You have a set amount of time to answer the questions. Don't get bogged down trying to answer a single question at the expense of 10 questions you can more readily answer.

COMPETENCY 1.0   SCIENCE METHODOLOGY, TECHNIQUES, AND HISTORY

Skill 1.1   Nature of scientific knowledge, inquiry, and historical perspectives: scientific methods and processes; facts, models, theories and laws; historical roots of science, and contributions made by major historical figures

The Scientific Method
The scientific method is the basic process behind scientific experimentation. It involves several steps, beginning with formulating a hypothesis and working through the discovery process to make a conclusion based on observation and testing.

Posing a question
Although many discoveries happen by chance, a scientist's standard thought process begins with forming a question to test by conducting research. The more limited the question, the more readily an experiment can be designed to answer that question.

Forming a hypothesis
Once the question is formulated, a scientist makes an educated guess about the answer to the problem or question. This 'best guess' is your hypothesis.

Doing the test
Next, a series of steps known as an experiment are outlined to test this hypothesis. To make a test fair, data from an experiment must have a **variable** or any condition that can be changed, for example temperature or mass. A good test will try to manipulate as few variables as possible. This allows the researcher to more readily identify the variable or condition that produces a particular result. Experiments also require a second factor known as a **control**. A control is a factor that remains unchanged throughout the experiment, which allows the researcher to verify that the experiment worked correctly. When using a control, all the conditions are the same except for the variable being tested.

Observing and recording the data
Once the experiment is conducted, data must be gathered based on the results obtained. Data reporting should state specifics of how the measurements were made during the experiment. For example, a graduated cylinder needs to be read with proper procedures. For beginning students, technique must be part of the instructional process so as to give validity to the data.

GENERAL SCIENCE   1

Drawing a conclusion
Careful analysis of the recorded data allows the experimenter to draw a conclusion based on the evidence. After recording data, compare your data with that of other researchers that conducted similar experiments. A conclusion is the judgment derived from the data results.

Scientific findings are usually documented in the form of a lab report. All lab reports should include a specific **title** and tell exactly what is being studied. The **abstract** is a summary of the report that is placed at the beginning of the paper. The **purpose** should always be defined, clearly stating the question the experiment was designed to answer. The purpose should include the **hypothesis** (educated guess) of the expected outcome of the experiment. The entire experiment should relate to this problem. It is important to accurately describe what was done to prove or disprove a hypothesis. A **control** is needed in every experiment; it is necessary to prove that the results obtained are a result of the manipulated variable. Only one variable should be manipulated at a time. **Observations** and **results** of an experiment should be recorded, including all results from data. Drawings, graphs, and illustrations should be included to support information. Observations are objective, whereas analysis and interpretation is subjective. A **conclusion** should explain why the results of the experiment either proved or disproved the hypothesis.

Scientific theory and experimentation must be repeatable. It is also possible that previously established theories can be disproved and may be changed on the basis of new scientific proof. Science depends on communication, agreement, and disagreement among scientists. It is built on theories, laws, and hypotheses.

Theory - the formation of principles or relationships which have been verified and accepted, a proven hypothesis

Law - an explanation of events that occur with uniformity under the same conditions (laws of thermodynamics, law of gravitation)

Hypothesis - an unproven theory or educated guess followed by research to best explain a phenomena

A **model** is a basic element of the scientific method. Many things in science are studied with models. A model is a simplification or representation of a problem that is being studied or predicted. A model is a substitute, but it is similar to what it represents. We encounter models at every step of our daily living. The periodic table of the elements is a model chemists use for predicting the properties of the elements. Physicists use Newton's laws to predict how objects will interact, such as planets and spaceships. In geology, the continental drift model estimates the past positions of continents. Samples, ideas, and methods are all examples of models. At every step of scientific study, models are extensively used. The primary activity of hundreds of thousands of U.S. scientists is to produce new models; these models are presented to the scientific community and the general public in tens of thousands of scientific papers published every year.

Historical perspective
Beginnings of Microbiology- Anton van Leeuwenhoek is known as the father of microscopy. In the 1650s, Leeuwenhoek made tiny lenses that produced magnifications up to 300x. He was the first scientist to see and describe bacteria, yeast, plants, and microscopic organisms found in water. Over the years, light microscopes have been refined to produce greater clarity and magnification. The scanning electron microscope (SEM) was developed in the 1950s. Instead of light, a beam of electrons is passed through the specimen. Scanning electron microscopes have a resolution about one thousand times greater than that of light microscopes. The disadvantage of the SEM is that the chemical and physical methods used to prepare the samples result in the death of the specimen.

In the late 1800s, Louis Pasteur discovered that microorganisms play a causal role in the onset of disease. He also pioneered the pasteurization process, and the development of the rabies vaccine. Robert Koch took Pasteur's observation that microorganisms cause disease one step further by postulating that specific diseases were caused by specific pathogens. **Koch's postulates, as his discoveries are called,** are still used as guidelines in the field of microbiology. The postulates state that: The same pathogen must be found in every person with the same disease, the pathogen must be isolated and grown in culture, when the organism is re-introduced into an experimental animal, that animal should develop the same disease originally seen, and the same pathogen must be isolated from the re-infected, experimental animal.

Discovery of DNA- DNA structure was another key discovery in biological study. In the 1950s, James Watson and Francis Crick discovered that the DNA molecule was organized into a double helix. The discovery of this structure made it possible to explain DNA's ability to replicate and to control protein synthesis..

Experimental Models-The use of animals in biological research has expedited many scientific discoveries. Animal research has allowed scientists to learn more about biological systems, including the circulatory and reproductive systems.

Animal models are used in a variety of applications including drug testing, vaccines development, and developing other products (such as perfumes and shampoos) before use or consumption by humans.

Skill 1.2   **Mathematics, measurement, and data manipulation; measurement and notation systems; data presentation and interpretation, including error analysis**

Science may be defined as a body of knowledge that is systematically derived from study, observations, and experimentation. Its goal is to identify and establish principles and theories which may be applied to solve problems. Pseudoscience, on the other hand, is not based on scientific methodology or application. Some of the more classic examples of pseudoscience include witchcraft, alien encounters, or topics that are explained by hearsay rather than by reproducible experimentation.

In Science, the metric system is the worldwide standard of measurement; this allows for easier comparison among experiments done by scientists around the world. Learn the following basic metric units and prefixes:

meter – base unit of length
liter – base unit of volume
gram – base unit of mass

deca-(meter, liter, gram)= 10X the base unit   **deci** = 1/10 the base unit
hecto-(meter, liter, gram)= 100X the base unit   **centi** = 1/100 the base unit
kilo-(meter, liter, gram) = 1000X the base unit   **milli** = 1/1000 the base unit

Graphing is an important way to visually display data for analysis. The two types of graphs most commonly used are the **line graph** and the **bar graph** (histogram). Line graphs are set up to show two variables, represented by one point on the graph. The *x*-axis is the horizontal axis; this is where the independent variable of the experiment is plotted. Independent variables are those that are not affected by any changes in the experimental conditions. A common example of an independent variable is time. Time proceeds regardless of any changes in experimental conditions. The *y*-axis is the vertical axis; the dependent variable is plotted here. Dependent variables are manipulated by the experimenter; factors such as the amount of light or the height of a plant are examples of dependent variables. The axes of a graph should be labeled at equal intervals. If one interval represents one day, the next interval should not represent ten days. A "best fit" line is drawn to join the points on a graph though it may connect all the points of the data. Both axes should always be labeled for a graph to be accurately interpreted. Graphs must always include a descriptive title; a good title will describe both the dependent and the independent variables. Bar graphs are set up with category labels along the horizontal axis. An appropriate scale is chosen for the vertical axis which will show the responding variable. A bar is drawn at each category with a height equal to the data along the vertical axis. Each bar represents a separate piece of data and is not joined by a continuous line. Bar graphs should also be given a descriptive title.

The type of graph used to represent one's data depends on the type of data collected. **Line graphs** are used to compare different sets of related data or to predict data that has not yet be measured. For example, a line graph would be used to compare the rate of activity of different enzymes at varying temperatures. A **bar graph** or **histogram** is used to compare different items and make comparisons based on this data. A bar graph would be used to compare the range of ages of children in a classroom. A **pie chart** is useful when organizing data as part of a whole. A pie chart would be used to display the percent of time students spend on various after-school activities.

Experimental Error
All experimental uncertainty is due to either random errors or systematic errors.

Random errors are statistical fluctuations in the measured data due to the precision limitations of the measurement device. Random errors usually result from the experimenter's inability to take the same measurement in exactly the same way to get exactly the same number.

Systematic errors, by contrast, are reproducible inaccuracies that are consistently made during an experiment at the same point. Systematic errors are often due to a problem which persists throughout the entire experiment.

Systematic and random errors refer to problems associated with making measurements. Mistakes made in the calculations or in reading the instrument are not considered in error analysis.

Skill 1.3 **Laboratory procedures and safety: techniques of safe preparation, storage, use, and disposal of laboratory and field materials; selection and use of appropriate laboratory equipment**

Dissections - Animals that are not obtained from recognized sources should not be used for laboratory experiments. Decaying animals or those of unknown origin may harbor pathogens and/or parasites that could be harmful to an experimenter's health. Specimens should be rinsed before handling. Non-latex gloves are desirable. If gloves are not available, students with sores or scratches should be excused from the activity. Formaldehyde is a carcinogen and should be avoided or disposed of according to district regulations. Students objecting to dissections for moral reasons should be given alternative assignments.

Live specimens - Biological experiments may be done with all animals except mammalian vertebrates and birds. Lower-order life forms and invertebrates may be used for experimentation. No physiological harm will be inflicted upon the animal. All animals housed and cared for in the school must be handled in a safe and humane manner. Animals are not to remain on school premises during extended vacations unless adequate care is provided. Many state laws stipulate that any instructor who intentionally refuses to comply with the laws may be suspended or dismissed.

Microbiology - Pathogenic organisms must never be used for experimentation. Students should adhere to the following rules at all times when working with microorganisms to avoid accidental contamination:

1. Treat all microorganisms as if they were pathogenic
2. Maintain sterile conditions at all times

If you are taking a national level exam you should check the Department of Education for your state mandated safety procedures. You will want to know what your state expects of you not only for the test but also for performance in the classroom and for the welfare of your students.

Laboratory Equipment

Bunsen burners - Hot plates should be used whenever possible to avoid the risk of burns or fire. If Bunsen burners are used, the following precautions should be followed:

1. Know the location of fire extinguishers and safety blankets, and train students in their use. Long hair and long sleeves should be secured and out of the way.

2. Turn the gas on slowly to about a quarter of maximum on and make a spark with the striker. The preferred method to light burners is to use strikers, rather than matches.

3. Adjust the air valve at the bottom of the Bunsen burner until the flame shows an inner cone.

4. Adjust the flow of gas to the desired flame height by using the adjustment valve.

5. Do not touch the barrel of the burner (it is hot).

Graduated Cylinders- These are used for precise measurements. They should always be placed on a flat surface. The surface of the liquid will form a meniscus (lens-shaped curve). The measurement is read at eye level, reading the <u>bottom</u> of this curve.

Balances - Electronic balances are easier to use but are more expensive. An electronic balance should always be used on a flat surface and tared (returned to zero) before measuring Substances should always be placed on a piece of weighing paper to avoid spills and/or damage to the instrument. Triple beam balances must be used on a level surface. There are screws located at the bottom of the balance to make any adjustments. Start with the largest counterweight first and proceed toward the last notch that does not tip the balance. Do the same with the next largest, etc until the pointer remains at zero. The total mass is the total of all the readings on the beams. Again, use weighing paper under the substance to protect the equipment.

Buret – A buret is used to dispense precisely measured volumes of liquid. A stopcock is used to control the volume of liquid being dispensed at a time.

Light microscopes are commonly used in laboratory experiments. Several procedures should be followed to properly care for this equipment:

- Clean all lenses with lens paper only.
- Carry microscopes with two hands; one on the arm and one on the base.
- Always begin focusing on low power, then switch to high power.
- Store microscopes with the low power objective down.
- Always use a cover slip when viewing wet mount slides.
- Bring the objective down to its lowest position, then adjust the fine focus, to avoid breaking the slide or scratching the lens.

Wet mount slides should be made by placing a drop of water on the specimen and then putting a glass cover slip on top of the drop of water. Placing the cover slip on the slide at a 45 degree angle will help avoid air bubbles. Total magnification is determined by multiplying the ocular (usually 10X) and the objective (usually 10X on low, 40X on high).

Laboratory Procedures

Chromatography uses the principles of capillary action to separate substances such as plant pigments. Molecules of a larger size will migrate up the paper more slowly, whereas smaller molecules will move more quickly and produce lines of pigments.

Spectrophotometry uses percent light absorbance to measure a color change, thus giving qualitative data a quantitative value.

Centrifugation is used to separate substances of varying densities, which is achieved by spinning substances at a high speed. The more dense part of a solution will sediment at the bottom of the test tube, while the lighter material will stay on top. For example, centrifugation is used to separate blood into blood cells and plasma, with the heavier blood cells settling to the bottom.

Electrophoresis uses electrical charges of molecules to separate them according to their size. The molecules, such as DNA or proteins, are pulled through a gel towards either the positive end of the gel box (if the material has a negative charge) or the negative end of the gel box (if the material has a positive charge).

Technology

Computer technology has greatly improved the collection and interpretation of scientific data. Molecular findings have been enhanced through the use of computer images. Technology has revolutionized access to data via the Internet and shared databases. The manipulation of data is enhanced by sophisticated software capabilities. Computer engineering advances have produced such products as MRIs and CT scans in medicine. Laser technology has numerous applications with refining precision.

Satellites have improved our ability to communicate and transmit radio and television signals. Navigational abilities have been greatly improved through the use of satellite signals.

Sonar technology uses sound waves to locate objects and is especially useful underwater. The sound waves bounce off the object and are used to assist in location.

Seismographs record vibrations in the earth and allow us to measure earthquake activity.

Using Laboratory Chemicals and Solutions

All laboratory solutions should be prepared as directed in the lab manual. Care should be taken to avoid contamination of solutions and samples. All glassware should be rinsed thoroughly with distilled water before use and cleaned well after use. Safety goggles should be worn while working with glassware, to protect the eyes. All solutions should be made with distilled water, since tap water contains dissolved particles that may affect the results of an experiment. Chemicals should be stored in a secured, dry area. Chemicals should be stored in accordance with the material safety data sheet. Acids are to be locked in a separate area. Used solutions should be disposed of according to local disposal procedures. Any questions regarding safe disposal or chemical safety may be directed to the local fire department.

COMPTENCY 2.0   THE PHYSICAL SCIENCES

Skill 2.1   Matter and energy: structure and properties of matter, occurrence and abundance of elements, physical and chemical changes, forms and transformations of energy, conservation of mass and energy

Matter and its Properties
Everything in our world is made up of **matter;** rocks, mushrooms, animals, and people are all made up of matter. Matter has two characteristics: It takes up space, and it has mass.

Mass is a measure of the amount of matter in an object. Two objects of equal mass will balance each other on a simple balance scale no matter where the scale is located. For instance, two rocks with the same mass that are in balance on Earth will also be in balance on the moon. They will feel heavier on Earth than on the moon because of the gravitational pull of the Earth.

Weight is the measure of the Earth's pull of gravity on an object. It can also be defined as the pull of gravity between two bodies. The units of weight measurement most commonly used are the pound (English measure) and the kilogram (metric measure). So, based on the previous example about mass, although the two rocks have the same mass, depending on where they are in the universe their weight may vary because of differences in gravitational pull.

In addition to mass, matter also has the property of volume. **Volume** is the amount of cubic space that an object occupies. Volume and mass together give a more exact description of an object. Two objects may have the same volume, but different masses, or the same mass but different volumes, etc. For instance, consider two cubes that are each one cubic centimeter, one made from plastic, one from lead. They have the same volume, but the lead cube has more mass. The measure that we use to describe the cubes takes into consideration both the mass and the volume. **Density** is the mass of a substance per unit of volume. If the density of an object is less than the density of a liquid, the object will float in the liquid. If the object is denser than the liquid, then the object will sink.

To find an object's density, you must measure its mass and its volume. Then divide the mass by the volume ($D = m/V$). Density is stated in grams per cubic centimeter (g/cm^3) where the gram is the standard unit of mass.

If the object is a regular shape, you can find the volume by multiplying the length, width, and height. However, if it is an irregular shape, you can find the volume by seeing how much water it displaces. Measure the volume of the water in the container before and after the object is submerged. The difference between the two volumes will be the volume of the object.

Specific gravity is the ratio of the density of a substance to the density of water. For instance, the specific density of one liter of turpentine is calculated by comparing its mass (0.81 kg) to the mass of one liter of water (1 kg):

$$\frac{\text{mass of 1 L turpentine}}{\text{mass of 1 L water}} = \frac{0.81 \text{ kg}}{1.00 \text{ kg}} = 0.81$$

Physical and Chemical Properties

Physical properties and chemical properties of matter describe the appearance and behavior of a substance respectively. A **physical property** can be observed without changing the identity of a substance. For instance, you can describe the color, mass, shape, and volume of a book. **Chemical properties** are properties of a substance that become apparent after a chemical reaction has occurred. The original substances change, or new substances are formed, as these chemical properties become apparent.. Baking powder goes through a chemical change as it changes into carbon dioxide gas during the baking process.

Matter can exist in different physical states. A **physical change** is a change that does not produce a new substance but changes the appearance of the substance. The freezing and melting of water is an example of a physical change. A **chemical change** (or chemical reaction) is any change of a substance into one or more different substances. When wood is burnt, it becomes ash; the chemical composition of ash and wood are different, therefore this represents a chemical change.

The Periodic Table

The periodic table tells us a lot about the various elements and their atomic structures. The periodic table was invented in the late 1800s by Dmitri Mendeleev, and he grouped all of the known elements according to the similarities of their characteristics. He wrote the law of chemical periodicity, which states that the properties of the various elements are functions of the atomic number of the element. Elements are arranged according to different groups and "periods" and are listed in order of their atomic number. The atomic number is equal to the number of protons in each atom. Usually, the number of protons is the same as the number of electrons.

In a typical periodic table, the atomic number is located below the symbol of the element. The atomic mass is located above the symbol. The atomic mass is listed in units called atomic mass units, where 1 AMU equals 1/12 times the mass of carbon in grams. This takes into account both the neutrons and protons in the element. Elements that share the same atomic number but have a different atomic mass are called **isotopes** of one another.

The figure below represents the basic unit of the periodic table:

```
┌─────────────────────┐
│                     │
│    Atomic Mass      │
│                     │
│                     │
│      SYMBOL         │
│                     │
│                     │
│   Atomic Number     │
│                     │
└─────────────────────┘
```

The periodic table is divided into "groups" and "periods". Groups represent the vertical columns that include elements that are similar with respect to their chemical and physical properties. The columns start with metals and progress to nonmetals. The groups are shown as below:

- **Group 1A:** Alkali metals; very reactive; never found free in nature; react readily with water; e.g.- sodium
- **Group 2A:** Alkaline earth elements; all are metals; occur only in compounds; react with oxygen in the general formula XO (where O is oxygen and X is Group 2A element); e.g.- magnesium
- **Group 3A:** Metalloids; includes aluminum (the most abundant metal on Earth); react with oxygen in the general formula X_2O_3
- **Group 4A:** Includes metals and nonmetals; nonmetals are at the top of the column and metals are at the bottom; react with oxygen in the general formula XO_2; e.g.- carbon
- **Group 5A:** All elements form an oxygen or sulfur compound with X_2O_3 or X_2S_3 formulas; e.g.- nitrogen
- **Group 6A:** Includes oxygen
- **Group 7A:** Elements combine violently with alkali metals to form salts; all are highly reactive; includes fluorine and chlorine
- **Group 8A:** Nobel gasses; not abundant on the earth; not reactive with other elements; e.g.- neon and argon.

Periodis are horizontal rows in the periodic table. The elements within a period have very different properties. The patterns that are displayed across a period are repeated as one moves to the next period however.

Work and Energy
Energy is the ability to do work or supply heat. Work is the transfer of energy to move an object a certain distance. It is motion against an opposing force. Lifting a chair into the air is work; the opposing force is gravity. Pushing a chair across the floor is work; the opposing force is friction.

According to the **First Law of Thermodynamics**: Energy, is neither created nor destroyed, rather it is converted from one form of energy to another. This also means that energy is neither created nor destroyed in ordinary physical and chemical processes (non-nuclear). Energy, in all of its forms must be conserved. The following equation reflects the aforementioned statement: In any system, $\Delta E = q + w$ (Δ = change, E = energy, q = heat and w = work). This means that the change in energy is always equal to the energy used plus the work done.

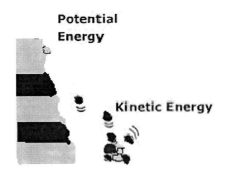

Potential Energy

Kinetic Energy

The two most commonly encountered forms of energy are potential and kinetic energy. **Kinetic energy** is the energy of a moving object. Potential energy is the energy stored in matter due to its position relative to other objects.

In any object, solid, liquid, or gas, the atoms and molecules that make up the object are constantly moving and colliding with each other. They are not stationary.

Due to this motion, the object's particles have varying amounts of kinetic energy. A fast moving atom can push a slower moving atom during a collision, so it has energy. All moving objects have energy, and that energy depends on the object's mass and velocity. Kinetic energy is calculated: $KE = \frac{1}{2} mv^2$.

An object's temperature is proportional to the average kinetic energy of the particles in the substance. When the temperature of a substance is increased its particles move faster, so their average kinetic energies increase as well. Temperature is *not* energy, it is not conserved.

The energy an object has due to its position or arrangement of its parts is called **potential energy**. Potential energy due to position is equal to the mass of the object times the gravitational pull on the object times the height of the object, or:

$$PE = mgh$$

Where PE = potential energy, m = mass of object, g = gravity, and h = height.

Heat

Heat is energy that is transferred between objects caused by differences in their temperatures. Heat is transferred from an object of higher temperature to one of lower temperature. This transfer continues until both objects reach the same temperature. Both kinetic energy and potential energy can be transformed into heat energy. When you step on the brakes in your car, the kinetic energy of the car is changed to heat energy by friction between the brake and the wheels. Other transformations can occur from kinetic to potential as well. Since most of the energy in our world is in a form that is not easily used, both man and nature have developed some clever ways of changing one form of energy into another form that may be more readily used

Skill 2.2 Heat and thermodynamics: thermal energy, measurement, transfer and effects on matter, first and second laws of thermodynamics

Heat and temperature are different physical quantities. **Heat** is a measure of energy. **Temperature** is a measure of the heat of an object.

Two concepts that are important in the discussion of temperature changes are thermal contact and thermal equilibrium. Objects are in **thermal contact** if they can affect each other's temperatures. Set a hot cup of coffee on a desk top. The two objects are in thermal contact with each other and will begin affecting each other's temperatures. The coffee will become cooler and the desktop warmer. Eventually, they will have the same temperature. When this happens, they are in **thermal equilibrium.**

Temperature

We cannot rely on our sense of touch to determine an object's temperature because it is not an accurate measurement. **Thermometers** are used to measure temperature. When a thermometer is used, a small amount of mercury, or colored alcohol as is more commonly used today, in a capillary tube will expand when heated. The metal end of the thermometer and the object whose temperature it is measuring are put in contact long enough for them to reach thermal equilibrium. Then the temperature can be read from the thermometer scale.

Three temperature units are used:

Celsius: The freezing point of water is set at 0 and the boiling point is 100 degrees. The interval between the two is divided into 100 equal parts called degrees Celsius.

Fahrenheit: The freezing point of water is 32 degrees and the boiling point is 212. The interval between is divided into 180 equal parts called degrees Fahrenheit.

Temperature readings can be converted from one to the other as follows.

Fahrenheit to Celsius	Celsius to Fahrenheit
C = 5/9 (F - 32)	F = (9/5) C + 32

Kelvin Scale has degrees the same size as the Celsius scale, but the zero point is moved to down to a hypothetical absolute zero. This is determined by the triple point of water. Water inside a closed vessel is in thermal equilibrium in all three states (ice, water, and vapor) at 273.15 Kelvin. This temperature is equivalent to .01 degrees Celsius. Because the degrees are the same in the two scales, temperature changes are the same in Celsius and Kelvin.

Temperature readings can be converted from Celsius to Kelvin:

Celsius to Kelvin	Kelvin to Celsius
K = C + 273.15	C = K - 273.15

The heat capacity of an object is the amount of heat energy that it takes to raise the temperature of the object by one degree.

Heat capacity (C) per unit mass (m) is called **specific heat** (c):

$$c = \frac{C}{m} = \frac{Q/\Delta}{m}$$

A **calorimeter** uses the transfer of heat from one substance to another to determine the specific heat of the substance. Specific heats for many materials have been calculated and can be found in tables.

There are a number of ways that heat is measured. In each case, the measurement is dependent upon raising the temperature of a specific amount of water by a specific amount. These conversions of heat energy and work are called the **mechanical equivalent of heat**.

The **calorie** is the amount of energy that it takes to raise one gram of water one degree Celsius.

The **kilocalorie** is the amount of energy that it takes to raise one kilogram of water by one degree Celsius. Food "calories" are actually kilocalories.

In the International System of Units **(SI),** the calorie is equal to 4.184 **joules.**

A **British thermal unit (BTU)** = 252 calories = 1.054 kJ

Heat energy that is transferred into or out of a system is **heat transfer.** The temperature change is positive for a gain in heat energy and negative when heat is removed from the object or system.

The formula for heat transfer is $Q = mc\Delta T$ where Q is the amount of heat energy transferred, m is the amount of substance (in kilograms), c is the specific heat of the substance, and ΔT is the change in temperature of the substance. It is important to assume that the objects in thermal contact are isolated and insulated from their surroundings.

Because energy is neither created nor destroyed, if a substance in a closed container loses heat, then another substance in the container must gain heat.

When an object undergoes a change of phase, it goes from one physical state (solid, liquid, or gas) to another. For instance, water can go from liquid to solid (freezing) or from liquid to gas (boiling). The heat that is required to change from one state to the other is called **latent heat.**

The **heat of fusion** is the amount of heat that it takes to change a substance from a solid to a liquid or the amount of heat released during the change from liquid to solid.

The **heat of vaporization** is the amount of heat that it takes to change a liquid to a gas.

Heat Transfer
Heat is transferred in three ways: **conduction, convection,** and **radiation.**

Conduction is the transfer of heat with no actual transfer of matter. Conduction takes place between two substances in contact with each other, or within one substance. Conduction is the movement of thermal energy (heat) between molecules that are in contact with each other.

The transfer rate is the ratio of the amount of heat per amount of time it takes to transfer heat from one area of an object to another. For example, if you place an iron pan on a flame, the handle will eventually become hot. How fast the handle gets hot is a function of the amount of heat and how long it is applied. The shorter the amount of time it takes to heat the handle, the greater the transfer rate.

Convection is the transfer of thermal energy within a fluid. The particles in a fluid (which could be air or a liquid) transfer the thermal energy from hot areas to cooler areas. Hot fluids rise and cooler fluids sink due to differences in their densities. This is the basic premise behind the transfer of heat energy by convection. The warmed, rising air from a heat source such as a fire or electric heater is a common example of convection. Convection ovens make use of circulating air to more efficiently cook food.

Radiation is the transfer of energy by waves such as the electromagnetic waves emitted by stars. The sun warms the earth by emitting radiant energy.

Thermodynamics

The relationship between heat and other forms of energy are the **laws of thermodynamics.**

The first law of thermodynamics is sometimes known as the law of conservation of energy. This law states that energy is conserved. This means that if energy is added to a system one of two things must happen. Either the thermal energy of the system must increase. Or work must be done. But that energy added to the system must be accounted for in some way - it is conserved. The change in heat energy supplied to a system (Q) is equal to the sum of the change in the internal energy (U) and the change in the work done by the system against internal forces. $\Delta Q = \Delta U + \Delta W$

The second law of thermodynamics says that thermal energy can move from colder system to warmer systems, only if work is done. It is intuitive that heat will move from a warm object to a cooler object. But is it is important to recognize that it can move the other way as well, with a little help. Heat cannot spontaneously pass from a colder to a hotter object. An ice cube sitting on a hot sidewalk will melt into a little puddle; but a puddle of water will not spontaneously cool and form the same ice cube.

Entropy is the measure of how much energy or heat is available for work. Work occurs only when heat is transferred from warmer to cooler objects. Once this is done, no more work can be done. The energy is still conserved, but is not available for work as long as the objects are the same temperature. According to this theory, all things in the universe will eventually reach the same temperature. If this happens, energy will no longer be usable.

Skill 2.3 Atomic and nuclear structure: atomic and nuclear structure and related chemical properties; nuclear transformations and characteristics of radioisotopes and radiation

An **atom** is the smallest particle of an element. Atoms consist of a nucleus surrounded by a cloud of moving electrons.
The **nucleus** is the center of the atom. There are two types of particles in the nucleus of an atom: protons and neutrons. **Protons** are small, positively charged particles. **Neutrons** are small uncharged particles in the nucleus. They are neutral. Neutrons and protons have about the same mass, but neutrons have no charge.

The **electrons** found moving outside the nucleus of an atom move rapidly in all direction. Electrons have a negative charge. Electrons are about 2,000 times less massive than protons.

The number of protons in the nucleus of an atom is called the **atomic number**. All atoms of the same element have the same atomic number. In an atom, the number of protons (with a positive charge) is equal to the number of electrons (with a negative charge).This makes an atom neutral. There are exceptions to this, but for now think of atoms as neutral.

The **atomic mass** of an atom is equal to the number of protons plus the number of neutrons. As mentioned before, the mass of an electron is so small, that their mass does not contribute considerably to the overall mass of an atom.

Isotopes
The number of protons of an atom is always the same. That is, an atom of carbon will always have 6 protons. But the number of neutrons can vary. For example, carbon can have 5, 6, 7,or even 8 neutrons. This means there are different values for the atomic mass of carbon. There are several different isotopes of carbon as a result. **Isotopes** of an element have the same number of protons in the nucleus, but have different masses. Neutrons explain the difference in mass. They have mass but no charge.

Scientists measure the mass of an atom by comparing it to the mass of carbon atom. Carbon has six protons and six neutrons and is called carbon-12. It is assigned a mass of 12 atomic mass units (amu), which is the standard unit for measuring the mass of an atom. Therefore, one amu is equal to 1/12 of the mass of a carbon atom.

The **mass number** of an atom is the sum of its protons and neutrons. In any element, there is a mixture of naturally occurring isotopes, some having slightly more or slightly fewer neutrons. The **atomic mass** of an element is an average of the mass numbers of its isotopes, based on the abundance of each isotope on Earth.

The following table summarizes the terms used to describe atomic nuclei.

Term	Example	Meaning	Characteristic
Atomic Number	# protons (p)	same for all atoms of a given element	Carbon (C) atomic number = 6 (6p)
Mass number	# protons + # neutrons (p + n)	changes for different isotopes of an element	C-12 (6p + 6n) C-13 (6p + 7n)
Atomic mass	average mass of the various isotopes of the element	usually not a whole number	atomic mass of carbon equals 12.011

More about electrons
Electrons orbit the nucleus of an atom in orbits.. Electrons orbiting the nucleus occupy energy levels that are arranged in order and the electrons tend to occupy the lowest energy level available. A **stable electron arrangement** is an atom that has all of its electrons in the lowest possible energy levels.

Each energy level holds a maximum number of electrons. However, an atom with more than one energy level does not hold more than 8 electrons in its outermost shell. The outermost electrons in the atoms are called **valence electrons.** These electrons are involved in the bonding process, and they determine the properties of the element.

Level	Name	Max. # of Electrons
First	K shell	2
Second	L shell	8
Third	M shell	18
Fourth	N shell	32

This can help explain why chemical reactions occur. Atoms react with each other when their outer levels are unfilled. When atoms either exchange or share electrons with each other, these energy levels become filled, and the atom becomes more stable.

As an electron gains energy, it moves from one energy level to a higher energy level. The electron can not leave one level until it has enough energy to reach the next level. **Excited electrons** are electrons that have absorbed energy and have moved farther from the nucleus.

Electrons can also lose energy. When they do, they fall to a lower energy level. However, they can only fall to the lowest level that has room for additional electrons.

Chemical Bonds

Chemical reactions involve the breaking and forming of bonds between atoms. Bonds involve only the outer electrons and do not affect the nucleus. When a reaction involves the nucleus, elements become different elements. This is called a **nuclear reaction**.

Binding energy is released when the nuclei of atoms are split apart in a nuclear reaction. This binding energy is called **nuclear energy**.

There are two types of nuclear reactions:

Nuclear fission occurs when atomic nuclei are split apart. Smaller nuclei are formed, and energy is released. The fission of many atoms in a short time period releases a large amount of energy. "Heavy water" is used in a nuclear reactor to slow down neutrons, thus controlling and moderating the nuclear reactions. A controlled environment that slowly releases energy allows for production of usable sources of energy such us nuclear submarines and nuclear power plants.

Nuclear fusion is the opposite of nuclear fission. It occurs when small nuclei combine to form a larger nucleus. It begins with the hydrogen atom which has the smallest nuclei. During one type of fusion reaction, four hydrogen nuclei are fused at very high pressures and temperatures. They form one helium atom. The sun and stars are examples of fusion. They are made mostly of hydrogen that is constantly fusing. As the hydrogen forms helium, it releases the energy that we see as light. When all of the hydrogen is used, the star will no longer shine. Scientists estimate that the sun has enough hydrogen to keep it glowing for another four billion years.

During a nuclear reaction, elements change into other elements called **radioactive elements**. Uranium is a radioactive element. The element uranium breaks down and changes into the element lead. Most natural radioactive elements breakdown slowly, so energy is released over a long period of time.

Radioactive particles are used in the treatment of cancer because they can kill cancer cells. However, if they are powerful enough, they can also be harmful to healthy cells. People working around such substances must protect themselves with the correct clothing, equipment, and procedures.

Skill 2.4	Mechanics: straight-line, projectile, circular, and periodic motion; Newton's laws of motion; work, energy, and power; simple machines; torque; friction; conservation of energy and momentum; gravity; Archimedes' principle and Bernoulli's principle

Dynamics is the study of the relationship between motion and the forces affecting motion. **Force** causes motion.

Mass and weight are not the same qualities. An object's **mass** gives it a resistance to change its current state of motion. It is also the measure of an object's resistance to acceleration. The force that the Earth's gravity exerts on an object with a specific mass is the object's weight on Earth. Weight is a force that is measured in Newtons. Weight (W) = mass times acceleration due to gravity (**W = mg**). To illustrate the difference between mass and weight, picture two rocks of equal mass on a balance scale. If the scale is balanced in one place, it will be balanced everywhere, regardless of the gravitational field.

However, the weight of the stones would vary on a spring scale, depending upon the gravitational field. In other words, the stones would be balanced both on earth and on the moon. However, the weight of the stones would be greater on earth than on the moon.

Newton's laws of motion:

Newton's first law of motion is also called the law of inertia. It states that an object at rest will remain at rest and an object in motion will remain in motion at a constant velocity unless acted upon by an external force.

Newton's second law of motion states that if a net force acts on an object, it will cause the acceleration of the object. The relationship between force and motion is force equals mass times acceleration. (**F = ma**).

Newton's third law of motion states that for every action there is an equal and opposite reaction. Therefore, if an object exerts a force on another object, that second object exerts an equal and opposite force on the first.

Machines

Simple machines include the following:

1. Inclined plane
2. Lever
3. Wheel and axle
4. Pulley

Compound machines are two or more simple machines working together. A wheelbarrow is an example of a complex machine. It uses a lever and a wheel and axle. Machines of all types ease workload by changing the size or direction of an applied force. The amount of effort saved when using simple or complex machines is called mechanical advantage or MA.

Work is done on an object when an applied force moves across a distance.

Power is the work done divided by the amount of time that it took to do it. (Power = Work / time)

Motion and resistance to motion

Surfaces that touch each other have a certain resistance to motion. This resistance is **friction.**

1. The materials that make up the surfaces will determine the magnitude of the frictional force.
2. The frictional force is independent of the area of contact between the two surfaces.
3. The direction of the frictional force is opposite to the direction of motion.
4. The frictional force is proportional to the normal force between the two surfaces in contact.

Static friction describes the force of friction of two surfaces that are in contact but do not have any motion relative to each other, such as a block sitting on an inclined plane. **Kinetic friction** describes the force of friction of two surfaces in contact with each other when there is relative motion between the surfaces.

Push and pull –Pushing a vacuum cleaner or pulling a bowstring applies muscular force when the muscles expand and contract. Elastic force occurs when an object returns to its original shape (for example, when a bow is released).

Rubbing – Friction opposes the motion of one surface past another. Friction is common when slowing down a car or sledding down a hill.

Pull of gravity – The force of attraction between two objects. Gravity exists not only on Earth but also between planets as well as in black holes.

Forces on objects at rest – The formula F= m/a is shorthand for force equals mass over acceleration. An object will not move unless the force is strong enough to move the mass. Also, there can be opposing forces holding the object in place. For instance, a boat can potentially be forced to drift away by underwater currents but an equal and opposite force, a docking rope, keeps it tied to the dock.

Forces on a moving object - Inertia is the tendency of any object to resist a change in motion. An object at rest tends to stay at rest. An object that is moving tends to keep moving.

Inertia and circular motion – The centripetal force is provided by the high banking of the curved road and by friction between the wheels and the road.

The Law of **Conservation of Energy** states that energy may neither be created nor destroyed. Therefore, the sum of all energies in a system remains a constant.

Example:

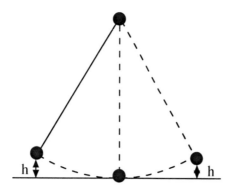

The formula to calculate the potential energy is PE = mgh.

The mass of the ball = 20kg
The height, h = 0.4m
The acceleration due to gravity, g = 9.8 m/s^2

PE = mgh
PE = 20(.4)(9.8)
PE = 78.4J (Joules, units of energy)

When the ball is to the extreme left, the Potential Energy (PE) = 78.4J resides while the Kinetic Energy (KE) = 0. As the ball approaches the center position, the PE decreases while the KE increases.. At exactly halfway between the left and center positions, the PE = KE.

The center position of the ball is where the Kinetic Energy is at its maximum while the Potential Energy (PE) = 0. At this point, theoretically, the entire PE has transformed into KE. Now the KE = 78.4J while the PE = 0.

As the ball swings to the extreme right, the Potential Energy (PE) is once again at its maximum and the Kinetic Energy (KE) = 0.

We can now say that:

PE + KE = 0
PE = -KE

The sum of PE and KE is the **total mechanical energy:**

Total Mechanical Energy = PE + KE

The law of **momentum conservation**: When two objects collide in an isolated system, the total momentum of the two objects before the collision is equal to the total momentum of the two objects after the collision. That is, the momentum lost by object 1 is equal to the momentum gained by object 2.

Straight-line, circular, and periodic motion

Matter can move in a straight line, in a circular pattern, and in a periodic fashion. The Greeks were the first recorded people to think about motion. They thought that matter wanted to be stopped. They were under the impression that once an object moved, it would not keep moving. They thought that the object would slow down and stop because its nature was to be at rest. These early scientists consider that matter moves. Galileo was the first to realize the error in the early scientists' thought process. Galileo concluded that an object keeps moving, even against the force of gravity..

Straight-line motion

To make an object move, a force must be applied. Friction must also be taken into account; it makes moving objects slow down. This characteristic was also noticed for the first time by Galileo. This is Newton's first law of motion, which states that an object at rest remains at rest unless acted upon by force. Force can have varied effects on moving objects. Force makes objects move, slow down, stop them, increase their speed, decrease their speed, etc.

A moving object has speed, velocity, and acceleration. To summarize, when force is applied to an object it moves in a straight line (Newton's first law) and adding force can make it go faster or slow it down.

Circular motion

Circular motion is defined as acceleration along a circle, a circular path or a circular orbit. Circular motion involves acceleration of the moving object by a centripetal force that pulls the moving object towards the center of the circular orbit. Without this acceleration, the object would move in a straight line, according to Newton's first law of motion. Circular motion is accelerated even though the speed is constant, because the object's velocity is constantly changing direction.

Let's look at some examples of circular motion. An artificial satellite orbiting the earth in a geosynchronous orbit; a stone which is tied to a rope and is being swung in circles; a race car turning through a curve in a racetrack; an electron moving perpendicular to a uniform magnetic field; a gear turning inside a mechanism.

A special kind of circular motion occurs when an object rotates around its own center of mass. The rotation around a fixed axis of a three-dimensional body involves the circular motion of its parts. This can be called spinning (or rotational) motion. When an object moves in a circular path, a force must be directed toward the center of the circle in order to keep the object moving. This constraining force is called **centripetal force**. Gravity is the centripetal force that keeps a satellite orbiting the earth.

Periodic motion

Periodic motion occurs when an object moves back and forth in a regular motion. Some examples of periodic motion are: weight on a string swinging back and forth (pendulum); a ball bouncing up and down. Periodic motion is characterized by three things:

-Velocity - they all have velocity, the bouncing ball, weight on a pendulum etc.
-Period - the period is the time the object takes to go back and forth. The time the ball takes to bounce back can be measured. Sometimes, the word period is replaced by the word frequency. Frequency is the reciprocal of period.
-Amplitude - the amplitude is half the distance the object goes from one side of the period to the other (the height of the pendulum or bouncing ball). When an object is rotating, the amplitude is the radius of the circle (1/2 the diameter).

There are many devices that use the characteristics of periodic motion. A clock is the most common example of periodic motion. Periodic motion is always used in the study of wave motion, including light, sound, and music.

Archimedes' Principle

This principle, which helps to explain the phenomena of buoyancy. is named after Archimedes of Syracuse, the Greek man who first discovered the relationship. Archimedes' principle states that **the buoyant force is equal to the weight of the displaced fluid.** The weight of the displaced fluid is directly proportional to the volume of the displaced fluid (if the surrounding fluid is of uniform density). Thus, among objects with equal masses, the one with greater volume has greater buoyancy.

Suppose a rock's weight is measured as 10 Newtons, when suspended by a string in a vacuum. When the rock is lowered by the string into the water, it displaces water weighing 3 Newtons. The buoyant force is 10 - 3 = 7 Newtons. The density of the immersed object relative to the density of the fluid is easily calculated without measuring any volumes:

$$\text{Relative density} = \frac{\text{Weight}}{\text{Weight - Apparent Immersed Weight}}$$

The applications of Archimedes' principle are many and important:
* Submarines
* Diving Weighting System
* Naval Architecture
* Flotation
* Buoyancy Compensator, and many more.

Bernoulli's Principle

Bernoulli's principle states that as the speed of a fluid, gas, or liquid increases, the pressure it exerts decreases. This principle is named for the mathematician and scientist, Daniel Bernoulli, though it was previously understood by Leonhard Euler and others. In a fluid flow with no viscosity, and therefore, one in which pressure difference is the only accelerating force, it is equivalent to Newton's laws of motion.

Bernoulli's principle also describes the venturi effect that is used in carburetors and other systems. In a carburetor, air is passed through a venturi tube in order to decrease its pressure. This happens because the air velocity has to increase as it flows through the constriction. Thus, Bernoulli's principle is of great practical application in aircraft flight and carburetors.

Skill 2.5	Electricity and magnetism: Characteristics of static and current electricity, electrical circuits, alternating and direct currents, transformers and motors, sources of EMF, magnetism

The electromagnetic spectrum consists of frequency (f), measured in hertz, and wavelength (λ), measured in meters. The frequency times the wavelength of every electromagnetic wave equals the speed of light (3.0×10^9 meters/second).

Roughly, the range of wavelengths in the electromagnetic spectrum is:

	$\underline{f}$		λ	
Radio waves	$10^5 - 10^{-1}$	hertz	$10^3 - 10^9$	meters
Microwaves	$10^{-1} - 10^{-3}$	hertz	$10^9 - 10^{11}$	meters
Infrared radiation	$10^{-3} - 10^{-6}$	hertz	$10^{11.2} - 10^{14.3}$	meters
Visible light	$10^{-6.2} - 10^{-6.9}$	hertz	$10^{14.3} - 10^{15}$	meters
Ultraviolet radiation	$10^{-7} - 10^{-9}$	hertz	$10^{15} - 10^{17.2}$	meters
X-Rays	$10^{-9} - 10^{-11}$	hertz	$10^{17.2} - 10^{19}$	meters
Gamma Rays	$10^{-11} - 10^{-15}$	hertz	$10^{19} - 10^{23.25}$	meters

Electricity

Electrostatics is the study of stationary electric charges. A plastic rod that is rubbed with fur or a glass rod that is rubbed with silk will become electrically charged and will attract small pieces of paper. The charge on the plastic rod rubbed with fur is negative and the charge on glass rod rubbed with silk is positive.

Electrically charged objects share these characteristics:

1. Like charges repel one another.
2. Opposite charges attract each other.
3. Charge is conserved. A neutral object has no net charge. If the plastic rod and fur are initially neutral, when the rod becomes charged by the fur a negative charge is transferred from the fur to the rod. The net negative charge on the rod is equal to the net positive charge on the fur.

Materials through which electric charges can easily flow are called conductors. On the other hand, an **insulator** is a material through which electric charges do not move easily, if at all.

A simple device used to indicate the existence of a positive or negative charge is called an **electroscope**. An electroscope is made up of a conducting knob, and attached to it are very lightweight conducting leaves usually made of foil (gold or aluminum). When a charged object touches the knob, the leaves push away from each other because like charges repel each other. It is not possible to tell whether the charge is positive or negative.

Charging by induction:

If you touch a knob with your finger while a charged rod is nearby, the electrons will be repulsed and flow out of the electroscope through the hand. If the hand is removed while the charged rod remains close, the electroscope will retain the charge.

When an object is rubbed with a charged rod, the object will take on the same charge as the rod. However, charging by induction gives the object the opposite charge as that of the charged rod.

Grounding charge:

Charge can be removed from an object by connecting it to the earth through a conductor. The removal of static electricity by conduction is called **grounding**.

An **electric circuit** is a path along which electrons flow. A simple circuit can be created with a dry cell, wire, and a bell or light bulb. When all are connected, the electrons flow from the negative terminal, through the wire to the device and back to the positive terminal of the dry cell. If there are no breaks in the circuit, the device will work; the circuit is closed. Any break in the flow will create an open circuit and cause the device to shut off.

The device (bell or bulb) is an example of a **load**. A load is a device that uses energy. Suppose that you also add a buzzer so that the bell rings when you press the buzzer. The buzzer is acting as a **switch**. A switch is a device that opens or closes a circuit. Pressing the buzzer makes the connection complete and the bell rings. When the buzzer is not engaged, the circuit is open and the bell is silent.

A **series circuit** is one where the electrons have only one path along which they can move. When one load in a series circuit goes out, the circuit is open. An example of this is a set of Christmas tree lights that is missing a bulb. None of the bulbs will work if one bulb is not working.

A **parallel circuit** is one where the electrons have more than one path to travel along. If a load goes out in a parallel circuit, the other load will continue to work because the electrons can still find a way to continue moving along the path.

When an electron goes through a load, it does work and therefore loses some of its energy. The measure of how much energy is lost is called the **potential difference**. The potential difference between two points is the work needed to move an electron from one point to another.

Potential difference is measured in a unit called the volt. **Voltage** is potential difference. The higher the voltage, the more energy the electrons have. This energy is measured by a device called a voltmeter. To use a voltmeter, place it in a circuit parallel with the load you are measuring.

Current is the number of electrons per second that flow past a point in a circuit. Current is measured with a device called an ammeter. To use an ammeter, put it in series with the load you are measuring.

As electrons flow through a wire, they lose potential energy. Some of the energy is changed to heat energy because of resistance. **Resistance** is the ability of the material to oppose the flow of electrons through it. All substances have some resistance, even if they are a good conductor, such as copper. This resistance is measured in units called **ohms**. A thin wire will have more resistance than a thick one because it will have less room for electrons to travel. In a thicker wire, there will be more possible paths for the electrons to flow. Resistance also depends upon the length of the wire. The longer the wire, the more resistance it will have.

Potential difference, resistance, and current form a relationship know as **Ohm's Law**. Current **(I)** is measured in amperes and is equal to potential difference **(V)** divided by resistance **(R)**.

$$I = V / R$$

If you have a wire with resistance of 5 Ohms and a potential difference of 75 volts, you can calculate the current by:

I = 75 volts / 5 ohms
I = 15 amperes

A current of 10 or more amperes will cause a wire to get hot. Twenty-two amperes is about the maximum for a house circuit. Current above 25 amperes can start a fire.

Electricity can also be used to change the chemical composition of a material. For instance, when electricity is passed through water, it breaks the water down into hydrogen gas and oxygen gas.

Circuit breakers in a home monitor the electric current. If there is a circuit overload, the circuit breaker will create an open circuit, stopping the flow of electricity.

Computers can be made small enough to fit inside a plastic credit card by creating what is known as a solid state device. In this device, electrons flow through solid material such as silicon.

Resistors are used to regulate volume on a television or radio; they are also used in dimmer switches for lights.

A bird can sit on an electrical wire without being electrocuted because the bird and the wire have about the same electrical potential. However, if that same bird would touch two wires at the same time, he would not have to worry about flying south next year.

When caught in an electrical storm, a car is a relatively safe place from lightening because of the resistance of the rubber tires. A metal building would not be safe unless there was a lightening rod that would attract the lightening and conduct it into the ground.

Magnetism
Magnets have a north pole and a south pole. Like poles repel and opposing poles attract. A **magnetic field** is the space around a magnet where its force affects objects. The closer you are to a magnet, the stronger the force. As you move away, the force becomes weaker.

Some materials act as magnets and some do not. This is because magnetism is a result of electrons in motion. The most important motion in this case is the spinning of the individual electrons. Electrons spin in pairs in opposite directions in most atoms. The magnetic field that each spinning electron creates is canceled by another electron spinning in the opposite direction.

In an atom of iron, there are four unpaired electrons. The magnetic fields of these are not canceled out. Their fields add up to make a tiny magnet. Their fields exert forces on each other setting up small areas in the iron called **magnetic domains** where atomic magnetic fields line up in the same direction.

You can make a magnet out of an iron nail by stroking the nail in the same direction repeatedly with a magnet. This causes poles in the nail to be attracted to the magnet. The tiny magnetic fields in the nail line up in the direction of the magnet. The magnet causes the domains pointing in its direction to grow in the nail. Eventually, one large domain results and the nail becomes a magnet.

A bar magnet has a north pole and a south pole. If you divide the magnet in half, it will have a north and south pole.

The earth has a magnetic field. In a compass, a tiny, lightweight magnet is suspended and will line its south pole up with the North Pole magnet of the Earth.

A magnet can be made out of a coil of wire by connecting the ends of the coil to a battery. When the current goes through the wire, the wire acts in the same way that a magnet does; it is called an **electromagnet**. The poles of the electromagnet will depend upon which way the electric current runs. An electromagnet can be made more powerful in three ways:

1. Make more coils.
2. Put an iron core (nail) inside the coils.
3. Use more battery power.

Telegraphs use electromagnets to work. When a telegraph key is pushed, current flows through a circuit, turning on an electromagnet that attracts an iron bar. The iron bar hits a sounding board that responds with a click. When the key is released, the electromagnet turns off. Messages can be sent around the world in this way.

Scrap metal can be removed from waste materials by the use of a large electromagnet that is suspended from a crane. When the electromagnet is turned on, the metal in the pile of waste will be attracted to it. All other materials will remain on the ground.

An **electric meter**, such as the one found on the side of a house, contains an aluminum disk that sits directly in a magnetic field created by electricity flowing through a conductor. The more the electricity flows (current), the stronger the magnetic field. The stronger the magnetic field, the faster the disk turns. The disk is connected to a series of gears that turn a dial. Meter readers record the number from that dial.

Air conditioners, vacuum cleaners, and washing machines use electric motors. An electric motor uses an electromagnet to change electric energy into mechanical energy.

In a **motor**, electricity is used to create magnetic fields that oppose each other and cause the rotor to move. The wiring loops attached to the rotating shaft have a magnetic field opposing the magnetic field caused by the wiring in the housing of the motor that cannot move. The repelling action of the opposing magnetic fields turns the rotor.

A **generator** is a device that turns rotary, mechanical energy into electrical energy. The process is based on the relationship between magnetism and electricity. As a wire, or any other conductor, moves across a magnetic field, an electric current occurs in the wire. The large generators used by electric companies have a stationary conductor; inside a magnet attached to the end of a rotating shaft is positioned inside a stationary conducting ring that is wrapped with a long, continuous piece of wire. When the magnet rotates, it induces a small electric current in each section of wire as it passes. Each section of wire is a small, separate electric conductor. All the small currents of these individual sections add up to one large current, which is used for electric power.

A **transformer** is an electrical device that changes electricity of one voltage into another voltage, usually from high to low. You can see transformers at the top of utility poles. It uses two properties of electricity: first, magnetism surrounds an electric circuit, and second, voltage is made when a magnetic field moves or changes strength. Voltage is a measure of the strength or amount of electrons flowing through a wire. If another wire is close to an electric current changing strength, the electric current will also flow into that other wire, as the magnetism changes. A transformer takes in electricity at a higher voltage and lets it run through many coils wound around an iron core. An output wire with fewer coils is also around the core. The changing magnetism makes a current in the output wire. Having fewer coils means less voltage, so the voltage is reduced.

Common sources of **EMFs** (electromagnetic fields) include power lines, appliances, medical equipment, cellular phones, and computers.

Skill 2.6 **Waves: characteristics of transverse and longitudinal waves; reflection, refraction, diffraction, and interference; Doppler effect; sound; electromagnetic radiation; color; optics**

Sound

Sound waves are produced by a vibrating body, which moves forward and compresses the air in front of it. It then reverses direction, so that the pressure on the air is lessened and expansion of the air molecules occurs. One compression and expansion creates one longitudinal wave. Sound can be transmitted through any gas, liquid, or solid. However, it cannot be transmitted through a vacuum, because there are no particles present to vibrate and bump into their adjacent particles to transmit the wave.

The vibrating air molecules move back and forth parallel to the direction of motion of the wave as they pass the energy from adjacent air molecules (closer to the source) to air molecules farther away from the source.

The **pitch** of a sound depends on the **frequency** that the ear receives. High-pitched sound waves have high frequencies. High notes are produced by an object that is vibrating at a greater rate per second than one that produces a low note.

The **intensity** of a sound is the amount of energy that crosses a unit of area in a given amount of time. The loudness of the sound is subjective and depends upon the effect on the human ear. Two tones of the same intensity but different pitches may appear to have different loudness. The intensity level of sound is measured in decibels. Normal conversation is about 60 decibels, while a power saw is about 110 decibels.

The **amplitude** of a sound wave determines its loudness, with loud sound waves creating larger amplitudes. The larger the sound wave, the more energy is needed to create the wave.

An **oscilloscope** is useful in studying waves because it gives a picture of the wave that shows the crest and trough of the wave. **Interference** is the interaction of two or more waves that meet. If the waves interfere constructively, the crest of each one meets the crests of the others. They combine into a crest with greater amplitude, creating a louder sound. If the waves interfere destructively, then the crest of one meets the trough of another. They produce a wave with lower amplitude that produces a softer sound.

If you have two tuning forks that produce different pitches, one will produce sounds of a slightly higher frequency. When you strike the two forks simultaneously, you will hear beats. **Beats** are a series of loud and soft sounds because when the waves meet, the crests combine at some points and produce loud sounds. At other points, they nearly cancel each other out and produce soft sounds.

When a piano tuner tunes a piano, he only uses one tuning fork, even though there are many strings on the piano. He adjusts the first string to be the same pitch as that of the tuning fork. Then he listens to the beats that occur when both the tuned and untuned strings are struck. He adjusts the untuned string until he can hear the correct number of beats per second. This process of striking the untuned and tuned strings together and timing the beats is repeated until all the piano strings are tuned.

Pleasant sounds have a regular wave pattern that is repeated over and over. Sounds that do not happen with regularity are unpleasant and are called **noise**.

Changes in experienced frequency due to relative motion of the source of the sound is called the **Doppler effect.** When a siren approaches, the pitch is high. When it passes, the pitch drops. As a moving sound source approaches a listener, the sound waves are closer together, causing an increase in frequency in the sound that is heard. As the source passes the listener, the waves spread out and the sound experienced by the listener is lower.

Waves

Transverse waves are characterized by the particle motion being perpendicular to the wave motion; **longitudinal waves** are characterized by the particle motion being parallel to the wave motion.

Transverse Wave

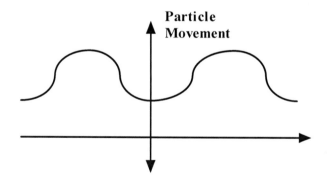

Direction of Energy Transport

Longitudinal Wave

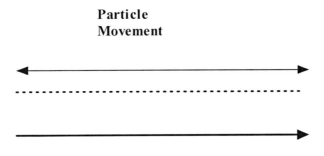

Wave **interference** occurs when two waves meet while traveling along the same medium. The medium takes on a shape resulting from the net effect of the individual waves upon the particles of the medium. There are two types of interference: constructive and destructive.

Constructive interference occurs when two crests or two troughs of the same shape meet. The medium will take on the shape of a crest or a trough with twice the amplitude of the two interfering crests or troughs. If a trough and a crest of the same shape meet, the two pulses will cancel each other out, and the medium will assume the equilibrium position. This is called **destructive interference**.

Destructive interference in sound waves will reduce the loudness of the sound. This is a disadvantage in rooms, such as auditoriums, where sound needs to be at its optimum. However, it can be used as an advantage in noise reduction systems. When two sound waves differing slightly in frequency are superimposed, beats are created by the alternation of constructive and destructive interference. The frequency of the beats is equal to the difference between the frequencies of the interfering sound waves.

Wave interference occurs with light waves in much the same manner that it does with sound waves. If two light waves of the same color, frequency, and amplitude are combined, the interference shows up as fringes of alternating light and dark bands. In order for this to happen, the light waves must come from the same source.

Light

When we refer to light, we are usually talking about a type of electromagnetic wave that stimulates the retina of the eye, or visible light. Each individual wavelength within the spectrum of visible light represents a particular **color**. When a particular wavelength strikes the retina, we perceive that color. The colors of visible light are sometimes referred to as ROYGBIV (red, orange, yellow, green, blue, indigo, and violet). The visible light spectrum ranges from red (the longest wavelength) to violet (the shortest wavelength) with a range of wavelengths in between. If all the wavelengths strike your eye at the same time, you will see white. Conversely, when no wavelengths strike your eye, you perceive black.

Shadows illustrate one of the basic properties of light. Light travels in a straight line, if you put your hand between a light source and a wall; you will interrupt the light and produce a shadow.

When light hits a surface, it is **reflected.** The angle of the incoming light (angle of incidence) is the same as the angle of the reflected light (angle of reflection). It is this reflected light that allows you to see objects. This happens when the reflected light reaches your eyes.

Different surfaces reflect light differently. Rough surfaces scatter light in many different directions. A smooth surface reflects the light in one direction. If it is smooth and shiny (like a mirror), you see your image in the surface.

When light enters a different medium, it bends. This bending, or change of speed, is called **refraction**.

Light can be **diffracted**, or bent around the edges of an object. Diffraction occurs when light goes through a narrow slit. As light passes through it, the light bends slightly around the edges of the slit. You can demonstrate this by pressing your thumb and forefinger together, making a very thin slit between them. Hold them about 8 cm from your eye and look at a distant source of light. The pattern you observe is caused by the diffraction of light.

Light and other electromagnetic radiation can be polarized because the waves are transverse. The distinguishing characteristic of transverse waves is that they are perpendicular to the direction of the motion of the wave. Polarized light has vibrations confined to a single plane that is perpendicular to the direction of motion. Light is able to be polarized by passing it through special filters that block all vibrations, except those in a single plane. Polarized sunglasses cut down on glare by blocking out all but one place of vibration..

Lens and Mirrors

The image that you see in a bathroom mirror is a virtual image because it only appears to be where it is. However, a curved mirror can produce a real image. A real image is produced when light passes through the point where the image appears. A real image can be projected onto a screen.

Cameras use a convex lens to produce an image on the film, which is thicker in the middle than at the edges. The image size depends upon the focal length (distance from the focus to the lens). The longer the focal length, the larger the image will be. A **converging lens** produces a real image whenever the object is far enough from the lens so that the rays of light from the object can hit the lens and be focused into a real image on the other side of the lens.

Eyeglasses can help correct deficiencies of sight by changing where the image seen is focused on the retina of the eye. If a person is nearsighted, the lens of his eye focuses images in front of the retina. In this case, the corrective lens placed in the eyeglasses will be concave so that the image will reach the retina. In the case of farsightedness, the lens of the eye focuses the image behind the retina. The correction will call for a convex lens to be fitted into the glass frames so that the image is brought forward into sharper focus.

Skill 2.7   Periodicity: the periodic table, trends in chemical and physical properties

The **periodic table of elements** is an arrangement of the elements in rows and columns so that it is easy to locate elements with similar properties. The elements of the modern periodic table are arranged in numerical order by atomic number.

The **periods** are the rows numbered down the left side of the table. They are called first period, second period, etc. The columns of the periodic table are called **groups**, or **families.** Elements in a family have similar properties.

There are three types of elements that are grouped by color: metals, nonmetals, and metalloids.

Element Key

Atomic
Number

↓

Electron → Arrangement **	2
	4 6
Symbol of → Element	**C**
	12.0 ←
	carbon

Average
Atomic
Mass

↑
Element
Name

** Number of electrons on each level. Top number represents the innermost level.

The periodic table arranges metals into families with similar properties. The periodic table has its columns marked IA - VIIIA, which are the traditional group numbers. Arabic numbers 1 - 18 are also used, as suggested by the Union of Physicists and Chemists. The Arabic numerals will be used in this text.

Metals:

With the exception of hydrogen, all elements in Group 1 are **alkali metals**. These metals are shiny, softer, and less dense than other metals and are the most chemically active.

Group 2 metals are the **alkaline earth metals.** They are harder, denser, have higher melting points, and are chemically active.

The **transition elements** can be found by finding the periods (rows) from 4 to 7 under the groups (columns) 3 - 12. They are metals that do not show a range of properties as you move across the chart. They are hard and have high melting points. Compounds of these elements are colorful, such as silver, gold, and mercury.

Elements can be combined to make other metallic objects. An **alloy** is a mixture of two or more elements having properties of metals. The elements do not have to be all metals. For instance, steel is made up of the metal iron and the non-metal carbon.

Nonmetals:

Nonmetals are not as easy to recognize as metals because they do not always share physical properties. However, in general, the properties of nonmetals are the opposite of metals. They are dull and brittle and are not good conductors of heat and electricity.

Nonmetals include solids, gases, and one liquid (bromine).

Nonmetals have four to eight electrons in their outermost energy levels and tend to attract electrons. As a result, the outer levels are usually filled with eight electrons. This difference in the number of electrons is what caused the differences between metals and nonmetals. The outstanding chemical property of nonmetals is that they react with metals.

The **halogens** can be found in Group 17. Halogens combine readily with metals to form salts. Table salt, fluoride toothpaste, and bleach all have an element from the halogen family.

The **Noble Gases** got their name because they did not react chemically with other elements, much like the nobility did not mix with the masses. These gases (found in Group 18) will only combine with other elements under very specific conditions. They are **inert** (inactive).

In recent years, scientists have found this to be only generally true, since chemists have been able to prepare compounds of krypton and xenon.

Metalloids:

Metalloids have properties in between metals and nonmetals. They can be found in Groups 13 - 16, but do not occupy the entire group. They are arranged in stair steps across the groups.

Physical Properties:
1. All are solids having the appearance of metals.
2. All are white or gray, but not shiny.
3. They will conduct electricity, but not as well as a metal.

Chemical Properties:
1. Have some characteristics of metals and nonmetals.
2. Properties do not follow patterns like metals and nonmetals. Each must be studied individually.

Boron is the first element in Group 13; it is a poor conductor of electricity at low temperatures. However, if its temperature increases it becomes a good conductor of electricity. By comparison, metals, which are good conductors, lose their ability as they are heated. It is because of this property that boron is so useful. Boron is a semiconductor. **Semiconductors** are used in electrical devices that have to function at temperatures too high for metals.

Silicon is the second element in Group 14; it is also a semiconductor and is found in great abundance in the earth's crust. Sand is made of a silicon compound, silicon dioxide. Silicon is also used in the manufacture of glass and cement.

Skill 2.8 **The mole and chemical bonding: the mole concept, the formulas and nomenclature of inorganic and simple organic compounds, bonding, electron dot and structural formulas, chemical composition and stoichiometry**

The mole is the base unit for measuring the amount of a substance. You may see its abbreviation, mol. Moles are determined for each substance by measuring the grams of the substance by a standard known as Avogadro's number (6.02×10^{23} atoms).

Organic Molecules

The hundreds of thousands of organic molecules have various chemical and physical properties and three-dimensional structures. However, certain similarities exist. Organic compounds are **covalently bonded, carbon-based molecules**. The ability of carbon atoms to bond with one another allows the formation of **long chains, double and triple bonds, and rings**.

Trends also exist in physical properties. Organic compounds tend to melt, boil, sublimate, or decompose below 300°C. Typically, they are highly flammable. Further, most organic molecules are only slightly soluble in water and dissolve better in organic solvents such as acetones or ethyl alcohol. However, **solubility and physical properties of organic compounds largely depend on their functional groups**. These **functional groups,** such as hydroxyl or amine groups, also determine the chemical properties of these molecules.

Many organic molecules contain functional groups, which are groups of atoms of a particular arrangement that gives the entire molecule certain characteristics. Functional groups are named according to their composition. The carboxyl group is the arrangement of -COOH atoms to make a molecule exhibiting acidic properties.

Some functional groups are polar and can ionize. For example, the hydrogen atom in the –COOH group can be removed (providing H^+ ions in solution). When this occurs, the oxygen retains both the electrons it shared with the hydrogen and will give the molecule a negative charge.

If polar or ionizing functional groups are attached to hydrophobic molecules, the molecule may become hydrophilic due to the functional group. Some ionizing functional groups are: -COOH, -OH, -CO, and $-NH_2$.

Some common functional groups include:

Hydroxyl group:

The hydroxyl group, -OH, is the functional group that identifies alcohols. The hydroxyl group makes the molecule polar, which increases the solubility of the compound.

Carbonyl group:

The carbonyl group is a -C=O attached to either a carbon chain or a hydrogen atom. It is found in aldehydes and ketones. If the carbon is bonded to a hydrogen, the molecule is an aldehyde.

Aldehyde

ketone

If the carbon is attached to two carbon chains, the molecule is a ketone.

The double-bonded oxygen atom is highly electronegative so it creates a polar molecule and will exhibit properties of polar molecules.

Carboxyl group:

The –COOH group has the ability to donate a proton or H^+ ion giving the molecule acidic properties.

Amino group:

An amino group contains an ammonia-like functional group composed of a nitrogen and two hydrogen atoms, covalently bonded. Since the nitrogen atom has unshared electrons, it can add H^+ ions (proton). This gives the molecule its basic properties.

An organic compound that contains an amino group is called an amine. The amines are weak bases because the unshared electron pair of the nitrogen atom can form a coordinate bond with a proton (H+ ion). Another molecule that contains an amino group is an amino acid. It consists of the –NH_2 group of an amine and the –COOH group of an acid.

H–CH–COOH   This is the amino acid glycine.
   |
   NH_2

Sulfhydryl group:

R——S——H

A **thiol** is a compound that contains a functional group that is composed of a sulfur atom and a hydrogen atom (-SH). This functional group is referred to as either a *thiol group* or a *sulfhydryl group*. Traditionally, thiols have been referred to as *mercaptans*.

The small difference in electronegativity between the sulfur and the hydrogen atom produce a non-polar, covalent bond. This, in turn, prevents hydrogen bonding, which gives thiols lower boiling points and less solubility in water than alcohols of a similar molecular mass.

Phosphate group:

$$\begin{array}{c} O \\ \| \\ -O-P-O \\ | \\ O \end{array}$$

The phosphate ion is contained in a hydrocarbon chain, making a phosphate group present in the molecule. This molecule is ideal for energy transfer reactions (ATP) because of its symmetry and rotating double bond.

In biological systems, phosphates are most commonly found in the form of adenosine phosphates (AMP, ADP and ATP), in DNA and RNA and can be released by the hydrolysis of ATP or ADP.

IUPAC is the International Union of Pure and Applied Chemistry. Organic compounds contain carbon and have their own branch of chemistry because of the huge number of carbon compounds in nature, including most molecules in living things. The simplest organic compounds are called hydrocarbons because they contain only carbon and hydrogen. Hydrocarbon molecules may be divided into classes of cyclic and open-chain compounds, depending on whether they contain a ring of carbon atoms. Open-chain molecules may be divided into branched or straight-chain categories.

Hydrocarbons are also divided into classes called **aliphatic** and **aromatic**. Aromatic hydrocarbons are related to benzene and always cyclic. Aliphatic hydrocarbons may be open-chain or cyclic. Aliphatic cyclic hydrocarbons are called **alicyclic**. Aliphatic hydrocarbons are separated into three groups: alkanes, alkenes, and alkynes.

Organic reactions tend to be complex processes that depend on relative electron affinity, bond strength, polarity and steric hindrance. Furthermore, reactions between organic molecules typically lead to reactive intermediates. The stability of such intermediates determines how and whether or not a reaction occurs.

A **chemical bond** is a force of attraction that holds atoms together. When atoms are bonded chemically, they give up their individual properties. For instance, hydrogen and oxygen combine to form water and no longer look like hydrogen and oxygen. They look like water.

A **covalent bond** is formed when two atoms share electrons. Recall that atoms whose outer shells are not filled with electrons are unstable. When they are unstable, they readily combine with other unstable atoms. By combining and sharing electrons, they act as a single unit. Covalent bonding happens among nonmetals. Covalent bonds are always polar between two non-identical atoms.

Covalent compounds are compounds whose atoms are joined by covalent bonds. Table sugar, methane, and ammonia are examples of covalent compounds.

An **ionic bond** is formed by the transfer of electrons. It happens when metals and nonmetals bond. Before chlorine and sodium combine, the sodium has one valence electron and chlorine has seven. Neither valence shell is filled, but the chlorine's valence shell is almost full. During the reaction, the sodium gives one valence electron to the chlorine atom. Both atoms then have filled shells and are stable. Something else happens during the bonding. Before the bonding, both atoms are neutral. When one electron is transferred, it upsets the balance of protons and electrons in each atom. The chlorine atom takes one extra electron, while the sodium atom releases one electron. The atoms now become **ions**– atoms with an unequal number of protons and electrons. To determine whether the ion is positive or negative, compare the number of protons (+ charge) to the electrons (- charge). If there are more electrons, the ion will be negative. If there are more protons, the ion will be positive. Therefore, the ionic bond is formed from the attraction of the positively charged ion to the negatively charged ion.

Compounds that result from the transfer of electrons from metal atoms to electrons from nonmetal atoms are called **ionic compounds.** Sodium chloride (table salt), sodium hydroxide (drain cleaner), and potassium chloride (salt substitute) are examples of ionic compounds.
Spontaneous diffusion occurs when random motion leads particles to increase entropy by equalizing concentrations. Particles tend to move into places of lower concentration. For example, sodium will move into a cell if the concentration is greater outside than inside the cell. Spontaneous diffusion keeps cells balanced.

Lewis dot structures are a method for keeping track of each atom's valence electrons in a molecule. Drawing Lewis structures is a three-step process:

1) Add the number of valence shell electrons for each atom. If the compound is an anion, add the charge of the ion to the total electron count because anions have more electrons than protons. If the compound is a cation, subtract the charge of the ion.
2) Write the symbols for each atom, showing how the atoms connect to each other.
3) Draw a single bond (one pair of electron dots or a line) between each pair of connected atoms. Place the remaining electrons around the atoms as unshared pairs. If every atom has an octet of electrons (except H atoms with two electrons), the Lewis structure is complete. Shared electrons count towards both atoms. If there are too few electron pairs to do this, draw multiple bonds (two or three pairs of electron dots between the atoms) until an octet is around each atom (except H atoms with two). If there are too many electron pairs to complete the octets with single bonds, then the octet rule is broken for this compound.

Example: Draw the Lewis structure of HCN.
Solution:

1) From their locations in the main group of the periodic table, we know that each atom contributes the following number of electrons: H—1, C—4, N—5. Because it is a neutral compound, the molecule will have a total of 10 valence electrons.

2) The atoms are connected with C at the center and will be drawn as: H C N.
   Having H as the central atom is impossible, because H has one valence electron and will always have only a single bond to one other atom. If N were the central atom, then the formula would probably be written as HNC.

3) Connecting the atoms with 10 electrons in single bonds gives the structure to the right. H has two electrons to fill its valence subshells, but C and N only have six each. A triple bond between these atoms fulfills the octet rule for C and N and is the correct Lewis structure.

$$H : \overset{..}{C} : \overset{..}{\underset{..}{N}}$$

$$H : C ::: N :$$

To select the most probable Lewis dot structure for a compound or molecule that follows the octet rule, review the structures and compare them to the method for constructing Lewis dot structures above.

Example: Which of the electron-dot structures given below for nitrous oxide (laughing gas), N_2O, are acceptable?

$$\text{I.} \quad : \overset{..}{N} : : \overset{..}{N} : : \overset{..}{O} :$$

$$\text{II.} \quad : \overset{..}{N} : \overset{..}{N} : : O :$$

$$\text{III.} \quad : N : : : \overset{..}{N} : : O :$$

Solution: Both nitrogen and oxygen follow the octet rule, so the Lewis structure should show each atom in the molecule with eight electrons, either unshared or shared. Upon examination, only choice I provides each atom in the molecule with eight electrons. Choice II has only six electrons around each of the nitrogen atoms, and choice III has 10 electrons around the center nitrogen atom.

There are chemical reactions occurring all around us everyday. So many different chemical reactions would be very difficult to understand. However, the millions of chemical reactions that take place each and every day fall into only a few basic categories. Using these categories can help predict products of reactions that are unfamiliar or new.

Once we have an idea of the **reaction type**, we can make a good prediction about the products of chemical equations, and also balance the reactions. **General reaction types** are listed in the following table. Some reaction types have multiple names.

Reaction type	General equation	Example
Combination Synthesis	$A + B \rightarrow C$	$2H_2 + O_2 \rightarrow 2H_2O$
Decomposition	$A \rightarrow B + C$	$2KClO_3 \rightarrow 2KCl + 3O_2$
Single substitution Single displacement Single replacement	$A + BC \rightarrow AB + B$	$Mg + 2HCl \rightarrow MgCl_2 + H_2$
Double substitution Double displacement Double replacement Ion exchange Metathesis	$AC + BD \rightarrow AD + BC$	$HCl + NaOH \rightarrow NaCl + H_2O$
Isomerization	$A \rightarrow A'$	

Chemical equations not only show the reactants and products, they also follow the law of conservation of mass, which states that matter cannot be created or destroyed, merely rearranged, in ordinary chemical reactions. Equations, then, must be *balanced*, to follow this law.

A properly written chemical equation must contain properly written formulas and must be **balanced**. Chemical equations are written to describe a certain number of moles of reactants becoming a certain number of moles of reaction products. The number of moles of each compound is indicated by its **stoichiometric coefficient** .

Example: In the reaction,

$$2H_2(g) + O_2(g) \rightarrow 2H_2O(l),$$

hydrogen has a stoichiometric coefficient of two, oxygen has a coefficient of one, and water has a coefficient of two because two moles of hydrogen react with one mole of oxygen to form two moles of water.

In a balanced equation, the stoichiometric coefficients are chosen such that the equation contains an **equal number of each type of atom on each side**. In our example, there are four H atoms and two O atoms on both sides. Therefore, the equation is properly written.

Balancing equations is a four-step process.

1) Write an **unbalanced equation**. This requires writing the chemical formulas for the species involved in the reaction.
2) Determine the **number of each type of atom on each side** of the equation to find if the equation is balanced.
3) Assume that **the molecule with the most atoms** has a stoichiometric coefficient of one, and determine the other stoichiometric coefficients required to create the **same number of atoms on each side** of the equation.
4) Multiply all the stoichiometric coefficients by a whole number if necessary to eliminate fractional coefficients.

Inorganic Molecules

Characteristics of inorganic compounds

Inorganic compounds do not contain carbon, as opposed to organic compounds. They include mineral salts, metals and their alloys, compounds of non-metallic elements such as phosphorous, and metal complexes. Because they encompass such a broad spectrum of compounds, it is difficult to group them. Some are naturally occurring minerals, and the vast majority of them are synthetic (man made).
Historically, inorganic compounds came principally from mineral sources of non-biological origin. Most known inorganic compounds are, however, synthetic and are not obtained directly from nature.

Many inorganic compounds exist in organisms and are essential. Sodium chloride and phosphorous ions are essential for life, as are some inorganic molecules, such as carbonic acid, nitrogen, carbon dioxide, and oxygen.

Nomenclature of inorganic compounds

There are two main types of inorganic compounds

1. Ionic inorganic compounds:

Some metals have only one valence electron (the electrons in the outermost energy level) e.g. sodium, lithium, aluminum, etc. These are found in groups I, II and III. These form positive ions by losing electrons.

Some metals have an unpredictable charge, though they form positive ions by losing electrons; these are called transition elements (e.g. Manganese, gold, iron, etc.). In cases where more that one electron is lost, the valence electron number is indicated in Roman numerals in parenthesis, e.g. Fe (II).

The elements and groups (in the case of polyatomic ions) with positive charges are written and named first. LiF is Lithium fluoride and NH_4Cl is Ammonium chloride. Metals, as a rule, form cations (positive ions).

The anions (negative ions) that are formed from single atoms are named by dropping the elemental suffix and adding "ide".

The overall charge of an ionic compound should be zero.

Examples of ionic inorganic compounds:

Fe_2O_3
$MgCl_2$
$CaCO_3$

2. Non-Ionic Inorganic compounds:

These are otherwise known as molecular inorganic compounds. These compounds have combinations of two or more non-metallic elements. Most of these compounds are combinations of elements from Groups IVA to VIIA with one another or with hydrogen.

When a hydrogen atom forms a compound with a nonmetal, the hydrogen atom is named first, and the nonmetal is named as if it were a negative ion (e.g. HF is hydrogen fluoride).

However, there are some exceptions to this rule. H_2O is called water, not dihydrogen oxide. When other elements combine to form binary compounds, the formula is usually written by putting the elements in order of increasing group number.

The number of atoms is given by one of the following prefixes: mono, di, tri, tetra, penta, hexa, hepta, and octa, etc. for numbers one through eight. The prefix, mono should not be used for the first element, and a prefix is always used for the second element.

Examples of molecular inorganic compounds:

SO_2	SO_3
NO_2	HCl
HBr	CO_2

Nomenclature rules:

1. Identify whether it is an ionic or a molecular compound.
If a metal or an ammonium ion is present, then it is an ionic compound.

2. If the compound is ionic, use the names and charges applicable to standard ionic compounds.

3. If the compound is a molecular compound, the rules for naming binary molecular compounds must be used.

Skill 2.9   **The kinetic theory and states of matter: kinetic molecular theory, phase characteristics and transformations, gas laws, characteristics of crystals**

The **phase of matter** (solid, liquid, or gas) is identified by its shape and volume. A **solid** has a definite shape and volume. A **liquid** has a definite volume, but no shape. A **gas** has no shape or volume because it will spread out to occupy the entire space of whatever container it is in.

Energy is the ability to cause change in matter. Applying heat to a frozen liquid changes it from solid to liquid. Continue heating it, and it will boil and vaporize, changing into a gas.

Evaporation is the change in phase from liquid to gas. **Condensation** is the change in phase from gas to liquid.

As a liquid is heated, the molecules begin moving faster within the container. As the substance becomes a gas and those molecules hit the sides of the container, pressure builds. **Pressure** is the force exerted on each unit of area of a surface. Pressure is measured in a unit called the **Pascal**. One Pascal (pa) is equal to one Newton of force pushing on one square meter of area.

Volume, temperature, and pressure of a gas are related to one another.

Temperature and pressure: As the temperature of a gas increases, its pressure increases. When you drive a car, the friction between the road and the tire, heats up the air inside the tire. Because the temperature increases, so does the pressure of the air on the inside of the tire. **Gay-Lussac's law** states that the pressure of a fixed amount of gas in a fixed volume is proportional to its temperature, or:

$$P \propto T$$

Or $P = kT$ where k is a constant. This gives a mathematical equation $\dfrac{P_1}{T_1} = \dfrac{P_2}{T_2}$ where the initial and final conditions are noted by the subscripts, 1 and 2.

Temperature and Volume: At a constant pressure, an increase in temperature causes an increase in the volume of a gas. **Charles's law** states that the volume of a fixed amount of gas at constant pressure is directly proportional to temperature, or:

$$V \propto T.$$

Or $V = kT$ where k is a constant. This gives a mathematical equation $\dfrac{V_1}{T_1} = \dfrac{V_2}{T_2}$.

These relations (pressure and temperature, and temperature and volume) are **direct variations**. As one component increases (decreases), the other also increases (decreases).

However, pressure and volume vary inversely, which means that as one characteristic increases, the other decreases.

Pressure and volume: At a constant temperature, a decrease in the volume of a gas causes an increase in its pressure. An example of this is a tire pump. The gas pressure inside the pump gets higher as you press down on the pump handle to compress the gas. This forces it to exist in a smaller volume. This relationship between pressure and volume is called **Boyle's Law. Boyle's law** states that the volume of a fixed amount of gas at constant temperature is inversely proportional to the gas pressure, or:

$$V \propto \frac{1}{P}.$$

Or $V = k/P$ where k is a constant. This gives a mathematical equation $P_1 V_1 = P_2 V_2$.

All of these relationships can be combined into the kinetic molecular theory. This theory explains that these three properties of a gas are based on the total effect of the individual molecules' motion and collision. This is also known as collision theory.

The behavior of an ideal gas can be determined as conditions, such as temperature, pressure, volume or quantity of gas—all of which can vary within a system. An **ideal gas** is an imaginary gas that obeys all of the assumptions of the kinetic molecular theory. While an ideal gas does not exist, most gases will behave like an ideal gas except when at very low temperatures or very high pressures.

The **combined gas law** uses the above laws to determine a proportionality expression that is used for a constant quantity of gas:

$$V \propto \frac{T}{P}.$$

The combined gas law is often expressed as an equality between identical amounts of an ideal gas at two different states:

$$\frac{P_1 V_1}{T_1} = \frac{P_2 V_2}{T_2}$$

Avogadro's hypothesis states that equal volumes of different gases at the same temperature and pressure contain equal numbers of molecules. It also states that that the volume of a gas at constant temperature and pressure is directly proportional to the quantity of gas, or:

$$V \propto n \text{ where } n \text{ is the number of moles of gas.}$$

Avogadro's law and the combined gas law yield $V \propto \frac{nT}{P}$. The proportionality constant, R, the **ideal gas constant,** is used to express this proportionality as the **ideal gas law**:

$$PV = nRT.$$

The ideal gas law ($PV = nRT$) is useful because it contains all the information of Charles's, Avogadro's, Boyle's, and the combined gas laws in a single expression.

Crystals

A crystal is a regular, repeating arrangement of atoms, ions or molecules. Crystals are well-organized structures.

There are two kinds of crystals, solid and liquid.

Solid crystals: As a liquid substance cools and forms a solid, molecules will arrange to form a solid with a repeating pattern, like the cubes of sodium chloride (salt). The strong attractive forces between oppositely charged ions, causes the repeating pattern of crystals and holds them together.

Liquid crystals: When certain solids melt, their crystal lattices disintegrate and their particles lose their three-dimensional pattern. However, when liquid crystals melt, they lose their rigid organization in only one or two dimensions. The inter-particle forces in a liquid crystal are relatively weak and therefore their arrangement is easily disrupted. When the lattice is broken, the crystal flows like a liquid. Liquid crystal displays (LCDs) are used in watches, thermometers, calculators and laptop computers because liquid crystals change with varying electric charge.

Characteristics of crystals:

1. Symmetry: Under certain operations, the crystal remains unchanged. The constituent atoms, molecules, or ions, are packed in a regularly ordered, repeating pattern extending in all three dimensions. Crystals form as they undergo a process of solidification; the result of solidification may be a single crystal or a group of crystals, a condition called a polycrystalline solid. The symmetry of crystals is one tool used in the classification of crystals.

2. Crystalline structures are universal: Crystalline structures occur in all classes of materials with ionic and covalent bonding. Sodium chloride is an example of a crystal formed out of ionic bonding. Graphite, diamond, and silica are examples of crystals with covalent bonding.

3. Crystallographic defects: Most crystalline structures have inborn imperfections. These defects have a great effect on the properties of the crystals.

4. Electrical properties: Some crystalline materials may exhibit special electrical properties such as the ferro-electric effect or the piezo-electric effect. Also, light passing through a crystal is often bent in different directions, producing an array of colors.

5. Crystal system: The crystal system is a grouping of crystal structures according to the axial system used to describe their lattice. Each crystal system consists of a set of three axes in a particular geometrical arrangement. There are seven unique crystal systems, of which the cubic is the most symmetrical. The other six (in decreasing order of symmetry) are: hexagonal, rhombohedral, orthorhombic, monoclinic and triclinic.

Skill 2.10   Chemical reactions: types of reactions; endothermic and exothermic reactions; effects of temperature, pressure, concentration, and presence of catalysts on reactions; practical applications of electrochemistry; balance chemical equations

A **chemical reaction** is a process where one or more molecules change into r, different molecules.

Also, energy is released during some chemical reactions. Sometimes the energy release is slow, sometimes it is rapid. In a fireworks display, energy is released very rapidly. However, the chemical reaction that produces tarnish on a silver spoon happens very slowly.

Reactants are the initial molecules in a reaction, and **products** are the molecules that result from the reaction. Chemical equilibrium is achieved when the quantities of reactants and products are at a 'steady state' and no longer shifting, though the reaction may still proceed forward and backward. The rate of forward reaction must equal the rate of backward reaction.

In one kind of chemical reaction, two elements combine to form a new substance. We can represent the reaction and the results in a chemical equation.

Carbon and oxygen form carbon dioxide. The equation can be written:

$$C \quad + \quad O_2 \quad \rightarrow \quad CO_2$$

1 atom of carbon	+	2 atoms of oxygen	$\rightarrow$ $\rightarrow$	1 molecule of carbon dioxide

No matter is ever gained or lost during a chemical reaction; therefore the chemical equation must be **balanced**. This means that there must be the same number of atoms on both sides of the equation. Remember that the subscript numbers indicate the number of atoms in the elements. If there is no subscript, assume there is only one atom.

In a second kind of chemical reaction, the molecules of a substance split and form two or more new substances. An electric current can split water molecules into hydrogen and oxygen gas.

$$2H_2O \quad \rightarrow \quad 2H_2 \quad + \quad O_2$$

2 molecules of water	$\rightarrow$ 2 molecules of hydrogen	+	1 molecule of oxygen

The number of molecules is shown by the number in front of an element or compound. If no number appears, assume that it is 1 molecule.

A third kind of chemical reaction is when elements change places with each other. An example of one element taking the place of another is when iron changes places with copper in the compound copper sulfate:

$$CuSo_4 \quad + \quad Fe \quad \rightarrow \quad FeSO_4 \; + \quad Cu$$

copper sulfate	+	iron (steel wool)	iron sulfate	copper

Sometimes two sets of elements change places. In this example, an acid and a base are combined:

$$HCl \quad + \quad NaOH \quad \rightarrow \quad NaCl \quad + \quad H_2O$$

| hydrochloric acid | sodium hydroxide | sodium chloride (table salt) | water |

Matter can change, but it cannot be created or destroyed. The sample equations show two things:

1. In a chemical reaction, matter is changed into one or more different kinds of matter.
2. The amount of matter present before and after the chemical reaction is the same.

Many chemical reactions give off energy. Like matter, energy can change form, but it can be neither created nor destroyed during a chemical reaction. This is the **law of conservation of energy.**

There are four kinds of chemical reactions:

In a **composition reaction**, two or more substances combine to form a compound.

$A + B \rightarrow AB$
i.e. silver and sulfur yield silver dioxide

In a **decomposition reaction**, a compound breaks down into two or more simpler substances.

$AB \rightarrow A + B$
i.e. water breaks down into hydrogen and oxygen

In a **single replacement reaction**, a free element replaces an element that is part of a compound.

$A + BX \rightarrow AX + B$
i.e. iron plus copper sulfate yields iron sulfate plus copper

In a **double replacement reaction**, parts of two compounds replace each other. In this case, the compounds switch partners.

AX + BY → AY + BX
i.e. sodium chloride plus mercury nitrate yields sodium nitrate plus mercury chloride

If, during a chemical reaction, more energy is needed to break the reactant bonds than released when product bonds form, the reaction is **endothermic** and heat is absorbed from the environment. The temperature of the environment goes down.

On the other hand, if more energy is released due to product bonds forming than is needed to break reactant bonds the energy is **exothermic** and the excess energy is released to the environment as heat. The temperature of the environment goes up.

The rate of most simple reactions **increases with temperature** because more molecules have the kinetic energy required to overcome the reaction's activation energy. The chart below shows the effect of temperature on the distribution of kinetic energies in a sample of molecules. These curves are called **Maxwell-Boltzmann distributions**. The shaded areas represent the amount of molecules containing sufficient kinetic energy for a reaction to occur. This area is larger at a higher temperature, so more molecules are above the activation energy and more molecules react per second.

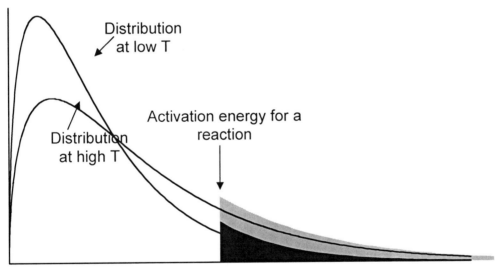

The website, http://www.mhhe.com/physsci/chemistry/essentialchemistry/flash/activa2.swf, provides an animated audio tutorial on energy diagrams.

Kinetic molecular theory may be applied to reaction rates in addition to physical constants like pressure. **Reaction rates increase with reactant concentration** because more reactant molecules are present, and more molecules are likely to collide with one another in a certain volume at higher concentrations. The nature of these relationships determines the rate law for the reaction. For ideal gases, the concentration of a reactant is its molar density, and this varies with pressure and temperature.

Kinetic molecular theory also predicts that **reaction rate constants increase with temperature** (values for k) because of two reasons:

1) More reactant molecules will collide with each other per second.
2) These collisions will each occur at a higher energy that is more likely to overcome the activation energy of the reaction.

A **catalyst** is a material that increases the rate of a chemical reaction without changing itself permanently in the process. Catalysts provide an alternate reaction mechanism for the reaction to proceed in both the forward and in the reverse direction. Therefore, **catalysts have no impact on the chemical equilibrium** of a reaction.

Catalysts reduce the activation energy of a reaction. This is the amount of energy needed for the reaction to begin. Molecules with such low energies that they would have taken a long time to react, will react more rapidly if a catalyst is present.

The impact of a catalyst may also be represented on an energy diagram. **A catalyst increases the rate of both the forward and reverse reactions by lowering the activation energy** for the reaction. Catalysts provide a different activated complex for the reaction at a lower energy state.

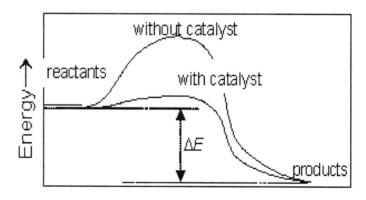

Reaction pathway→

Electrochemistry is the field that studies electrically driven reactions. A specific reaction that is driven by electricity is called an electrolysis reaction. An example would be adding electricity to split a water molecule into its components. It is also commonly used for the synthesis of organic molecules.

Practical applications for electrochemistry include pH measuring devices and electroplating. An example of an electrochemical device is a pH meter. They measure pH by comparing the electrical properties of a solution to those of a reference electrode. Electroplating is done by immersing a metal object in a bath containing the salt of a second metal, you can uniformly plate on a coating of that metal by running an electric current through the solution. The object being plated is called a cathode.

Skill 2.11   Solutions and solubility: types of solutions; solvents and the dissolving process; effects of temperature and pressure on solubility acids, bases, and salts; pH; buffers

A solution is a homogeneous mixture. Solutions are composed of a solvent and a solute. The solute is the dissolved particles in a solution. The solvent is the substance in which the solute is dissolved. For example, consider a glass of sugar water. The solvent in this solution is water. The solute is the sugar. The sugar is dissolved in the water to form a solution. Water is a very common solvent in many chemical reactions.
The solution process depends upon intermolecular forces.

Solutions form when the intermolecular forces between solute and solvent molecules are about as strong as those that exist in the solute alone or in the solvent alone. For example, $NaCl$ (salt) dissolves in water because:

1) The water molecules interact with the Na^+ and Cl^- ions with sufficient strength to overcome the attraction between them in the crystalline form.

2) Na^+ and Cl^- ions interact with the water molecules with sufficient strength to overcome the attraction water molecules have for each other in the liquid.

The intermolecular attraction between solute and solvent molecules is known as **solvation**. When the solvent is water, it is known as **hydration**. The figure to the left shows a hydrated Na^+ ion.

Polar and non-polar solutes and solvents

A non-polar liquid like heptane (C_7H_{16}) has intermolecular bonds with relatively weak London dispersion forces. Heptane is immiscible (not able to combine) in water because the attraction that water molecules have for each other via hydrogen bonding is too strong.

Unlike Na^+ and Cl^- ions, heptane molecules cannot reduce hydrogen bonding. Because bonds of similar strength must be broken and formed for solvation to occur, non-polar substances tend to be soluble in non-polar solvents, while ionic and polar substances are soluble in polar solvents like water. Polar molecules are often called **hydrophilic,** and non-polar molecules are called **hydrophobic.** This observation is often stated as "**like dissolves like.**" Network solids (e.g., diamond) are soluble in neither polar nor non-polar solvents because the covalent bonds within the solid are too strong for these solvents to break.

Temperature and solubility

When attempting to dissolve a solid like NaCl into a liquid, a higher temperature facilitates the solubility of the ionic solution, and more of the NaCl will enter into the solution. If the solution is highly saturated at a high temperature, the NaCl will fall out of solution and re-crystallize as a solid when the solution cools.

Solids in Liquids

The formula for solutions is: $CV = n$, where C is the molar concentration, V is the volume in liters of the liquid, and n is the number of moles of solute. Further, $n = m/Fw$, where m is the mass and Fw is the formula weight (or molar mass) of the solute. Solving for the mass, $m = C\, V\, Fw$.

How do you create a solution at a specific concentration? First, weigh the solid to obtain the mass. To determine the amount of solute needed, multiply the desired concentration by the volume of solution and by the formula weight of the solute. Place the mass of solute in a volume measuring device such as a volumetric flask or graduated cylinder. Add water to the volume desired and mix until dissolved.

The act of dissolving a solid into a liquid is a process that happens on the surface of the particles of the solute. The smaller the particles (the larger the surface area), the faster the solute dissolves. Confectioner's sugar has smaller particles than regular table sugar. Rock candy is just regular table sugar that has been crystallized in large lumps. When you put each crystal size of the chemically identical materials in your mouth, which one dissolves faster? The confectioner's sugar tastes sweetest because more of it has dissolved in the shortest amount of time. You can only taste dissolved sugar.

Expose the surface area of the solid to more solid and the solute will dissolve faster. Mixing also helps dissolve the solid. You can try this with sugar. Take two glasses of water at the same temperature, and add a spoonful of sugar to each. Mix one, but not the other. In which glass does the sugar dissolve more easily?

Most solid materials will dissolve faster with increased temperature. Since the increased temperature increases the motion of the molecules, you can think of this effect as being similar to mixing. For example, sugar dissolves more quickly in warm tea than iced tea. Table salt dissolves more quickly in hot water than in cold.

Electrolytes and precipitates

All NaCl molecules are present in solution as ions. Compounds that are completely ionized in water are called **strong electrolytes** because these solutions easily conduct electricity; most salts are strong electrolytes. Other compounds (including many acids and bases) may dissolve in water without completely ionizing. These are referred to as **weak electrolytes** because their state of ionization is at equilibrium with the larger molecule. Those compounds that dissolve with no ionization (e.g., glucose, $C_6H_{12}O_6$) are called **nonelectrolytes**

Particles in solution are free to move about and collide with each other, vastly increasing the likelihood that a reaction will occur, compared with solid particles. Aqueous solutions may react to produce an insoluble substance that will fall out of solution as a solid or gas **precipitate** in a **precipitation reaction**. An aqueous solution may also react to form **additional liquid solvent**.

Vapor pressure lowering, boiling point elevation, freezing point lowering

After a nonvolatile solute is added to a liquid solvent, only a fraction of the molecules at the surface of the liquid are now volatile and capable of escaping into the gas phase. The vapor consists of an essentially pure solvent that is able to condense freely. This imbalance drives equilibrium away from the vapor phase and into the liquid phase and **lowers the vapor pressure** by an amount proportional to the solute particles present.

It follows that from a lowered vapor pressure, a higher temperature is required for a vapor pressure equal to the external pressure over the liquid. Thus **the boiling point is raised** by an amount proportional to the solute particles present.

Solute particles dissolved in a liquid are not normally soluble in the solid phase of that solvent. When solvent crystals freeze, they typically align themselves with each other at first and keep the solute out. This means that only a fraction of the molecules, in the liquid at the liquid-solid interface, are capable of freezing, while the solid phase consists of essentially pure solvent that is able to melt freely. This imbalance drives equilibrium away from the solid phase and into the liquid phase and **lowers the freezing point** by an amount proportional to the solute particles present. Boiling point elevation and freezing point depression are both caused by a lower fraction of solvent molecules in the liquid phase than in the other phases.

Gases dissolved in liquids

Gases dissolved in liquids are easily measured when you know the pressure, volume, and temperature of the gas. Seltzer water and ammonia water are two good examples of solutions of a gas in a liquid. Seltzer, or carbonated water, consists of pressing the carbon dioxide gas into the water. The bubbles in beer or sparkling wines are also due to carbon dioxide, but the $CO2$ is a natural product of the fermentation process. Ammonia water, also called ammonium hydroxide solution, is made from ammonia ($NH3$) being pressed into water. It is used as a weak base and as a cleaning material.

Because the process is better done under pressure, it is often difficult to directly observe the actual dissolving process. The notable exception is the addition of dry ice, solid carbon dioxide, to water. As with a solid dissolving in a liquid, a gas dissolves in a liquid more easily with agitation or mixing. If you remove the carbonated beverage from its container, pressure is still necessary to keep the gas into solution. As the beverage sits for a few hours, the taste becomes what we describe as "flat" because, almost all of the carbon dioxide has escaped from the liquid. The only $CO2$ left in the water will produce a partial pressure equal to the partial pressure of the gas in the atmosphere. Water carries dissolved oxygen from the partial pressure of the oxygen in the atmosphere.

When the combination of liquid and gas is NOT at the lowest energy condition, an increase in temperature causes the separation. Lower temperature favors dissolving the gas into the liquid, while higher temperatures will cause the separation of the gas from the liquid.

COMPETENCY 3.0   THE LIFE SCIENCES

Skill 3.1   **The Cell: biologically important inorganic and organic molecules, structure and function of cells, cell organelles, cellular bioenergetics, the cell cycle and cytokinesis, meiosis and mitosis, homeostasis**

The structure of a cell is related to the cell's function. Root hair cells differ from flower stamens or leaf epidermal cells because they all have different functions.

Animal cells – begin a discussion of the nucleus as a round body inside the cell. It controls the cell's activities. The nuclear membrane contains threadlike structures called chromosomes. The genes are units that control cell activities found in the nucleus. The cytoplasm has many structures in it. Vacuoles are membranes that contain either food or waste materials. Animal cells differ from plant cells because they do not have a cell wall.

Plant cells – A unique feature of plant cells is their cell wall. A cell wall differs from cell membranes. The cell membrane is very thin and is a part of the cell, while the cell wall is thick and is a nonliving part of the cell. Chloroplasts are bundles of chlorophyll.

Photosynthesis - The process through which plants make use of energy that is produced by the sun to produce sugar. During photosynthesis, plants release oxygen into the air that is needed for us to breathe.

Single-celled organisms – A single-celled organism is called a **protist**. When you look under a microscope, the animal-like protists are called **protozoans.** They do not have chloroplasts and are classified by the way they move for food. Amoebas engulf other protists by flowing around and over them. The paramecium is covered by hair-like fibers, called cilia, which allows it to move back and forth searching for food. The euglena is a protozoan that moves with a tail-like structure called flagella.

Plant-like protists have cell walls and float in the ocean, of which **bacteria** are the simplest. A bacterial cell is surrounded by a cell wall, but there is no nucleus inside the cell. Most bacteria do not contain chlorophyll so they do not make their own food. The classification of bacteria is by shape. Cocci are round, bacilli are rod-shaped, and spirilla are spiral shaped.

Parts of Eukaryotic Cells

1. Nucleus - The "brain" of the cell which contains:

Chromosomes- DNA, RNA, and proteins tightly coiled to conserve space while providing a large surface area.
Chromatin – the loose structure of chromosomes. Chromosomes are called chromatin when the cell is not dividing.
Nucleoli – the place where ribosomes are made. These are seen as dark spots in the nucleus.
Nuclear membrane - contains pores, which let RNA out of the nucleus. The nuclear membrane is connected to the endoplasmic reticulum, which allows the membrane to expand or shrink if needed.

2. Ribosomes - the site of protein synthesis. Ribosomes may be free floating in the cytoplasm or attached to the endoplasmic reticulum. There may be up to a half million ribosomes in a cell, depending on how much protein is made by the cell.

3. Endoplasmic Reticulum - These are folded and provide a large surface area. They are the "roadway" of the cell and allow for transport of materials. The lumen of the endoplasmic reticulum helps to keep materials out of the cytoplasm and headed in the right direction. The endoplasmic reticulum is capable of building new membrane material. There are two types of endoplasmic reticulum:

Smooth Endoplasmic Reticulum - contain no ribosomes on their surface.

Rough Endoplasmic Reticulum - contain ribosomes on their surface. This form of endoplasmic reticulum is abundant in cells that make many proteins, like in the pancreas, which produces many digestive enzymes.

4. Golgi Complex or Golgi Apparatus - This structure is stacked to increase surface area. The Golgi Complex functions to sort, modify, and package molecules that are made in other parts of the cell. These molecules are either sent out of the cell or to other organelles within the cell.

5. Lysosomes – are found mainly in animal cells. These contain digestive enzymes that break down food, substances that are not needed, viruses, damaged cell components, and eventually the cell itself. It is believed that lysosomes are responsible for the aging process.

6. Mitochondria - large organelles that make ATP. Adenosine Tri-Phosphate is the molecule that stores energy that is harnessed during chemical reactions to break down ATP, thereby releasing energy in the cell. Muscle cells have many mitochondria because they use a great deal of energy. The folds inside the mitochondria are called cristae. They provide a large surface where the reactions of cellular respiration occur. Mitochondria have their own DNA and are capable of reproducing themselves if a greater demand is made for additional energy. Mitochondria are found only in animal cells.

7. Plastids – are only found in photosynthetic organisms. They are similar to the mitochondria due to their double membrane structure. They also have their own DNA and can reproduce if increased capture of sunlight becomes necessary. There are several types of plastids:

> **Chloroplasts** - green, function in photosynthesis. They are capable of trapping sunlight.
> **Chromoplasts** - make and store yellow and orange pigments; they provide color to leaves, flowers and fruits.
> **Amyloplasts** - store starch and are used as a food reserve. They are abundant in roots like potatoes.

8. Cell Wall - found in plant cells only, is composed of cellulose and fibers. It is thick enough for support and protection, yet porous enough to allow water and dissolved substances to enter. Cell walls are joined to each other.

9. Vacuoles - hold stored food and pigments. Vacuoles are very large in plants. This allows them to fill with water to provide turgor pressure. Lack of turgor pressure causes a plant to wilt.

10. Cytoskeleton - composed of protein, filaments attached to the plasma membrane and organelles. They provide a framework for the cell and aid in cell movement. They constantly change shape and move about. Three types of fibers make up the cytoskeleton:

> **Microtubules** - largest of the three; makes up cilia and flagella for locomotion. Flagella grow from a basal body, some examples are sperm cells, and tracheal cilia. Centrioles are also composed of microtubules. They form the spindle fibers that pull the cell apart into two cells during cell division. Centrioles are not found in the cells of higher plants.
> **Intermediate Filaments** - they are smaller than microtubules but larger than microfilaments. They help the cell to keep its shape.
> **Microfilaments** - smallest of the three, they are made of actin and small amounts of myosin (like in muscle cells). They function in cell movement such as cytoplasmic streaming, endocytosis, and ameboid movement.

The purpose of cell division is to provide growth, repair body (somatic) cells, and replenish or create sex cells for reproduction. There are two forms of cell division. Mitosis is the division of somatic cells and **meiosis** is the division of sex cells (eggs and sperm). The table below summarizes the major differences between the two processes.

Mitosis	Meiosis
1. Division of somatic cell	1. Division of sex cells
2. Two cells result from each division	2. Four cells or polar bodies result from each division
3. Chromosome number is identical to parent cells.	3. Chromosome number is half the number of parent cells
4. For cell growth and repair	4. Recombinations provide genetic diversity

Some terms to know:

Gamete - sex cell or germ cell; eggs and sperm.
Chromatin - loose chromosomes; this state is found when the cell is not dividing.
Chromosome - tightly coiled, visible chromatin; this state is found when the cell is dividing.
Homologues - chromosomes that contain the same information, are the same length, and contain the same genes.
Diploid - 2n number; diploid chromosomes are a pair of chromosomes (somatic cells).
Haploid - 1n number; haploid chromosomes are half of a pair (sex cells).

Mitosis

The cell cycle is the life cycle of the cell. It is divided into two stages; **interphase** and **mitotic division,** where the cell is actively dividing. Interphase is divided into three steps; G1 (growth) period, where the cell is growing and metabolizing, S period (synthesis) where new DNA and enzymes are being made, and the G2 phase (growth) where new proteins and organelles are being made to prepare for cell division. The mitotic stage consists of the stages of mitosis and the division of the cytoplasm.

The stages of growth and division are as follows. Be sure to know the correct order of steps.

1. Interphase - chromatin is loose, chromosomes are replicated, cell metabolism is occurring. Interphase is <u>not</u> a stage of mitosis.

2. Prophase - once the cell enters prophase, it proceeds through the following steps continuously, without stopping. The chromatin condenses to become visible chromosomes. The nucleolus disappears and the nuclear membrane breaks apart. Mitotic spindles form that will eventually pull the chromosomes apart; they are composed of microtubules. The cytoskeleton breaks down and the spindles are pushed to the poles, or opposite ends, of the cell by the action of centrioles.

3. Metaphase - kinetechore fibers attach to the chromosomes, which cause the chromosomes to line up in the center of the cell.

4. Anaphase - centromeres split in half, and homologous chromosomes separate. The chromosomes are pulled to the poles of the cell, with identical sets at either end.

5. Telophase - two nuclei form with a full set of DNA, each identical to the parent cell. The nucleoli become visible, and the nuclear membrane reassembles. A cell plate is visible in plant cells, whereas a cleavage furrow is formed in animal cells. The cell is pinched into two cells. Cytokinesis, or division, of the cytoplasm and organelles occurs.

Meiosis contains the same five stages as mitosis, but is repeated in order to reduce the chromosome number by one half. This way, when the sperm and egg join during fertilization, the haploid number is reached. The steps of meiosis are as follows:

Major function of Meiosis I - chromosomes are replicated; cells remain diploid.

Prophase I - replicated chromosomes condense and pair with homologues forming a tetrad. Crossing over (the exchange of genetic material between homologues to further increase diversity) occurs during Prophase I.
Metaphase I - homologous sets attach to spindle fibers after lining up in the middle of the cell.
Anaphase I - sister chromatids remain joined and move to the poles of the cell.
Telophase I - two new cells are formed, chromosome number is still diploid.
Cytokinesis- when cytokinesis begins, there are multiple nuclei within one cell. The purpose of cytokinesis is to organize each nucleus into its own individual cell. Cytokenesis begins in anaphase and continues on through telophase. Now the two cells will continue the cell cycle and begin their interphase again.

Major function of Meiosis II - to reduce the chromosome number in half

Prophase II - chromosomes condense.
Metaphase II - spindle fibers form again, sister chromatids line up in the center of cell, centromeres divide and sister chromatids separate.
Anaphase II - separated chromosomes move to opposite ends of cell.

Telophase II - four haploid cells form for each original sperm germ cell. One viable egg cell gets all the genetic information and three polar bodies form with no DNA. The nuclear membrane reforms and cytokinesis occurs.

Metabolism

Metabolism is the sum of all the chemical changes in a cell that convert nutrients to energy and macromolecules, the complex chemical molecules important to cell structure and function. The four main classes of macromolecules are **polysaccharides (carbohydrates), nucleic acids, proteins, and lipids**. Metabolism consists of two contrasting processes, anabolism and catabolism. Anabolism is biosynthesis, the formation of complex macromolecules from simple precursors. Anabolic reactions require the input of energy to proceed. Catabolism is the breaking down of macromolecules obtained from the environment or cellular reserves to produce energy in the form of ATP and basic precursor molecules. The energy produced by catabolic reactions drives the anabolic pathways of the cell.

Anabolism

The anabolic pathways of a cell diverge, synthesizing a large variety of macromolecules. All anabolic reactions produce complex molecules by linking small subunits, called monomers, together to form a large unit, or polymer. The main mechanisms of anabolism are condensation reactions that covalently link monomer units and release water.

Polysaccharides (carbohydrates) consist of monosaccharide units (e.g. glucose) linked together by glycosidic linkages, which are covalent bonds formed through condensation reactions. Glycogen is the principle storage form of glucose in animal and human cells. Cells produce glycogen by linking glucose monomers to form polymer chains.

Nucleic acids are large polymers of nucleotides. Cells link nucleotides through condensation reactions, consisting of a five-carbon sugar, a phosphate group, and a nitrogenous base. During DNA and RNA synthesis, the template molecule dictates the sequence of nucleotides by complementary base pairing.

Proteins are large polymers of amino acid subunits called polypeptides. Cells synthesize proteins by linking amino acids, forming peptide linkages through condensation reactions. RNA sequences direct the synthesis of proteins.

Lipids are a diverse group of molecules that are hydrophobic, or insoluble in water. Cells synthesize lipids from fatty acid chains formed by the addition of two-carbon units derived from a molecule called acetyl coenzyme A (acetyl-CoA). The reactions involved in lipid synthesis include condensation, oxidation/reduction, and alkylation.

Catabolism

The catabolic pathways of a cell break down macromolecules and produce energy to drive the anabolic pathways. In addition, catabolic pathways release precursor molecules (e.g. amino acids, nucleotides) used in biosynthesis. The basic reaction of catabolism is hydrolysis, the addition of a water molecule across a covalent bond.

Cells break the glycosidic linkages of stored or consumed polysaccharides, releasing glucose or other sugars that can be converted to glucose. The cells further degrade glucose to basic chemical end products, producing energy in the form of ATP.

Cells break down consumed proteins into amino acid units and other simple derivatives. Cells then use the amino acids to form new peptide chains or convert the derivative units into new amino acids. Cells can also acquire energy from the degradation of proteins, but the energy yield is not as high as that of polysaccharides and fatty acids.

Hydrolysis of lipids releases fatty acids that are a rich energy source. Fatty acids contain more than twice as much potential energy as do carbohydrates or proteins. The break down of fatty acids produces basic chemical compounds and energy in the form of ATP.

Finally, hydrolysis of nucleic acids by enzymes produces oligonucleotides (short strings of DNA or RNA) that are further degraded to produce free nucleosides (sugar-nitrogenous base units). Cells further digest nucleosides, separating the nitrogenous base from the sugar. Digestion of nucleosides ultimately results in the production of nitrogenous bases, simple sugars, and basic precursor compounds used in the synthesis of new DNA or RNA.

Biologically important organic molecules are involved in the processes of catabolism and anabolism. As described above, they include: polysaccharides, monosaccharides, nucleic acids, proteins, and lipids.

Plants require adequate amounts of nitrogen and phosphorus to build many cellular structures. The availability of the important inorganic minerals phosphorus and nitrogen often is the main limiting factor of biomass production.

Cellular bioenergetics

Cellular bioenergetics is the comparison of energy investment with the flow of energy through the cell. In this area of study, we ask if the product is worth the energy investment it requires. In the case of cells, we see that metabolism and reproduction are both worth the energy input because the cell is profiting in both areas. Photosynthesis results in usable energy, as does cellular respiration. It is important to note that optimal conditions will improve bioenergetics, and can sometimes have an affect on the cellular pathway chosen.

For example, let's look at bacteria and plants. Chemosynthetic bacteria are accustomed to living without light, so in lieu of solar energy, they oxidize sulfites or ammonia (chemosynthesis). In some cases, hydrogen atoms are present and the chemical pathway shifts. If hydrogen is available, the organism then takes advantage of the hydrogen. The energy produced from the reaction between hydrogen molecules and carbon dioxide is enough to fuel the production of biomass. If hydrogen is absent from the environment, the organism must use a different chemosynthetic pathway. It will utilize energy for chemosynthesis from reactions between O_2 and hydrogen sulfide or ammonia. In this scenario, the chemosynthetic microorganisms are dependent on photosynthesis, which occurs elsewhere, to produce the O_2 that they require. Thus, the environment causes a shift in pathways. The organism utilizes the pathway that is more beneficial (produces the most energy with the least work). Plants are similar in that they can function through either C3 or C4 photosynthesis. The key is their environment.

C3 photosynthesis is the typical mechanism of photosynthesis that most plants use. C4 photosynthesis, on the other hand, is an adaptation to arid environmental conditions because it results in more efficient use of water. Greater efficiency results in improved cellular bioenergetics.

Homeostasis
All living organisms respond and adapt to their environments. Homeostasis is the result of regulatory mechanisms that help maintain an organism's internal environment within tolerable limits. For example, in humans and mammals, constriction and dilation of blood vessels near the skin help maintain body temperature.

Skill 3.2 **Molecular basis of heredity and classical genetics: DNA replication, protein synthesis, Mendelian and non-Mendelian inheritance, mutations and transposable elements, genetic engineering, human genetic disorders, recombinant DNA, mapping the human genome, the interaction between heredity and the environment**

DNA and DNA REPLICATION

The modern definition of a gene is a unit of genetic information. DNA makes up genes which in turn make up the chromosomes. DNA is wound tightly around proteins in order to conserve space. The DNA/protein combination makes up the chromosome. DNA controls the synthesis of proteins, thereby controlling the total cell activity. DNA is capable of making copies of itself.

Review of DNA structure:

1. Made of nucleotides: a five carbon sugar, phosphate group and nitrogen base (either adenine, guanine, cytosine or thymine).

2. Consists of a sugar/phosphate backbone, which is covalently bonded. The bases are joined down the center of the molecule and are attached by hydrogen bonds which are easily broken during replication.

3. The amount of adenine (A) equals the amount of thymine (T) and the amount of cytosine (C) equals the amount of guanine (G).

4. The shape is that of a twisted ladder called a double helix. The sugar/phosphates make up the sides of the ladder and the base pairs make up the rungs of the ladder.

DNA Replication

Enzymes control each step of the replication of DNA. The molecule untwists, the hydrogen bonds between the bases break and serve as a pattern for replication. Free nucleotides found inside the nucleus join on to form a new strand. Two new pieces of DNA are formed which are identical; this is a very accurate process. There is only one mistake for every billion nucleotides added. This is because there are enzymes (polymerases) present that proofread the molecule. In eukaryotes, replication occurs in many places along the DNA at once. The molecule may open up at many places like a broken zipper. In prokaryotic circular plasmids, replication begins at a point on the plasmid and goes in both directions until it meets itself.

Base pairing rules are important in determining the sequence of a new strand of DNA. For example, our original strand of DNA has the following sequence:

1. A T C G G C A A T A G C This may be called our sense strand as it contains a sequence that makes sense or codes for something. The complementary strand (or other side of the ladder) would follow base pairing rules (A and T bond with each other and C and G bond with each other) and would read:

2. T A G C C G T T A T C G When the molecule opens up and nucleotides join on, the base pairing rules create two new identical strands of DNA.

1. A T C G G C A A T A G C   and   A T C G G C A A T A G C
   T A G C C G T T A T C G   2. T A G C C G T T A T C G

Protein Synthesis

It is necessary for cells to manufacture new proteins for growth and repair of the organism. Protein synthesis is the process that allows the DNA code to be read and carried out of the nucleus into the cytoplasm in the form of RNA. This is where the ribosomes are found, which are the sites of protein synthesis. The protein is then assembled according to the instructions on the DNA. There are several types of RNA. Familiarize yourself with where they are found and their function.

Messenger RNA - (mRNA) copies the code from DNA in the nucleus and takes it to the ribosomes in the cytoplasm.

Transfer RNA - (tRNA) free floating in the cytoplasm. Its job is to carry and position amino acids for assembly on the ribosome.

Ribosomal RNA - (rRNA) found in the ribosomes make a place for the proteins to be made. rRNA is believed to have many important functions, so much research is currently being done currently in this area.

Along with enzymes and amino acids, the RNA's function is to assist in the building of proteins. There are two stages of protein synthesis:

Transcription - This phase allows for the assembly of mRNA and occurs in the nucleus where the DNA is found. The DNA splits open, the mRNA reads the code and "transcribes" the sequence onto a single strand of mRNA. For example, if the code on the DNA is T A C C T C G T A C G A, the mRNA will make a complementary strand reading: A U G G A G C A U G C U (uracil replaces thymine in RNA). Each group of three bases is called a **codon**. The codon will eventually code for a specific amino acid to be carried to the ribosome.

"Start" codons begin the building of the protein and "stop" codons end transcription. When the stop codon is reached, the mRNA separates from the DNA and leaves the nucleus for the cytoplasm.

Translation - This is the assembly of the amino acids to build the protein and occurs in the cytoplasm. The nucleotide sequence is translated to choose the correct amino acid sequence. As the rRNA translates the code at the ribosome, tRNAs which contain an **anticodon**, seek out the correct amino acid and bring it back to the ribosome. For example, using the codon sequence from the example above:

The mRNA reads A U G / G A G / C A U / G C U
The anticodons are U A C / C U C / G U A / C G A
The amino acid sequence would be: Methionine (start) - Glu - His - Ala.

*Be sure to note if the table you are given is written according to the codon sequence or the anticodon sequence. It will be specified.

This whole process is accomplished through the assistance of **activating enzymes**. Each of the twenty amino acids has their own enzyme. The enzyme binds the amino acid to the tRNA. When the amino acids get close to each other on the ribosome, they bond together using peptide bonds. The start and stop codons are called nonsense codons. There is one start codon (AUG) and three stop codons. (UAA, UGA, and UAG). Addition mutations will cause the whole code to shift, thereby producing the wrong protein or, at times, no protein at all.

Mendelian and Non Mendelian Inheritance

Gregor Mendel is recognized as the father of genetics. His work in the late 1800s formed the basis of our knowledge of genetics. Although unaware of the presence of DNA or genes, Mendel realized there were factors (now known as genes) that were transferred from parents to their offspring. Mendel worked with pea plants and fertilized the plants himself, keeping track of subsequent generations which led to the Mendelian laws of genetics. Mendel found that two "factors" governed each trait, one from each parent. Traits, or characteristics, came in several forms, known as alleles. For example, the trait of flower color had white alleles and purple alleles. Mendel formed three laws:

> **Law of dominance** - in a pair of alleles, one trait may cover up the allele of the other trait. For example, brown eyes are dominant to blue eyes.

> **Law of segregation** - only one of the two possible alleles from each parent is passed on to the offspring from each parent. During meiosis, the haploid number insures that half the sex cells get one allele, half get the other.

Law of independent assortment - alleles sort independently of each other. Many combinations are possible, depending on which sperm fertilizes which egg. Compare this to the many combinations of hands possible when dealing a deck of cards.

monohybrid cross - a cross using only one trait.

dihybrid cross - a cross using two traits, and more combinations are possible.

Punnet squares - these are used to show the possible ways that genes combine, and indicate the probability of the occurrence of a certain genotype or phenotype. One parent's genes are put at the top of the box and the other parent at the side of the box. Genes combine on the square just like numbers that are added in addition tables we learned in elementary school.

Example: Monohybrid Cross - four possible gene combinations

Example: Dihybrid Cross - sixteen possible gene combinations

Some Definitions to Know

Dominant - the stronger of the two traits. If a dominant gene is present, it will be expressed. It is represented by a capital letter.

Recessive - the weaker of the two traits. In order for the recessive gene to be expressed, there must be two recessive genes present. It is represented by a lowercase letter.

Homozygous - (purebred) having two of the same genes present; an organism may be homozygous dominant with two dominant genes or homozygous recessive with two recessive genes.

Heterozygous - (hybrid) having one dominant gene and one recessive gene. The dominant gene will be expressed due to the Law of Dominance.

Genotype - the genes the organism has. Genes are represented with letters. AA, Bb, and tt are examples of genotypes.

Phenotype - how the trait is expressed in an organism. Blue eyes, brown hair, and red flowers are examples of phenotypes.

Incomplete dominance - neither gene masks the other; a new phenotype is formed. For example, red flowers and white flowers may have equal strength. A heterozygote (Rr) would have pink flowers. If a problem occurs with a third phenotype, incomplete dominance is occurring.

Co-dominance - genes may form new phenotypes. The ABO blood grouping is an example of co-dominance. A and B are of equal strength and O is recessive. Therefore, type A blood may have the genotypes of AA or AO, type B blood may have the genotypes of BB or BO, type AB blood has the genotype A and B, and type O blood has two recessive O genes.

Linkage - genes that are found on the same chromosome usually appear together unless crossing over has occurred in meiosis. (Example - blue eyes and blonde hair)

Lethal alleles - these are usually recessive due to the early death of the offspring. If a 2:1 ratio of alleles is found in offspring, a lethal gene combination is usually the reason. Some examples of lethal alleles include sickle cell anemia, Tay-Sachs and cystic fibrosis. Usually the coding for an important protein is affected.

Inborn errors of metabolism - these occur when the protein affected is an enzyme. Examples include PKU (phenylketonuria) and albinism.

Polygenic characters - many alleles code for a phenotype. There may be as many as twenty genes that code for skin color. This is why there is such a variety of skin tones. Another example is height; a couple of medium height may have very tall offspring.

Sex linked traits - the Y chromosome found only in males (XY) carries very little genetic information, whereas the X chromosome found in females (XX) carries very important information. Since men have no second X chromosome to cover up a recessive gene, the recessive trait is expressed more often in men. Women need the recessive gene on both X chromosomes to show the trait. Examples of sex linked traits include hemophilia and color-blindness.

Sex influenced traits - traits are influenced by the sex hormones. Male pattern baldness is an example of a sex influenced trait. Testosterone influences the expression of the gene; mostly men loose their hair due to this trait.

Sometimes DNA is not replicated perfectly. Inheritable changes in DNA are called **mutations**. Mutations may be errors in replication or a spontaneous rearrangement of one or more segments by factors like radioactivity, drugs, or chemicals. The amount of the change is not as critical as where the change is. Mutations may occur on somatic or sex cells, usually the ones on sex cells are more dangerous since they contain the basis of all information for the developing offspring. Mutations are not always bad; they are the basis of evolution, and if they make a more favorable variation that enhances the organism's survival, then they are beneficial. But, mutations may also lead to abnormalities, birth defects, and even death. There are several types of mutations:

Normal - A B C D E F

Duplication - one gene is repeated: A B C C D E F

Inversion - a segment of the sequence is flipped around: A E D C B F

Deletion - a gene is left out: A B C E F

Insertion or Translocation - a segment from another place on the DNA is inserted in the wrong place: A B C R S D E F

Breakage - a piece is lost: A B C (DEF is lost)

Non-disjunction – This occurs during meiosis when chromosomes fail to separate properly. One sex cell may get both genes and another may get none. Depending on the chromosomes involved, this may or may not be serious. Offspring may end up with either an extra chromosome or a missing one. An example of non-disjunction is Down Syndrome, where three of chromosome 21 are present.

Genetic Engineering

In its simplest form, genetic engineering requires enzymes to cut DNA, a vector, and a host organism for the recombinant DNA. A **restriction enzyme** is a bacterial enzyme that cuts foreign DNA in specific locations. The restriction fragment that results can be inserted into a bacterial plasmid (**vector**). Other vectors that may be used include viruses and bacteriophage. The splicing of restriction fragments into a plasmid results in a recombinant plasmid. This recombinant plasmid can now be placed in a host cell, usually a bacterial cell, to replicate.

The use of **recombinant DNA** provides a means to transplant genes among different species. This opens the door for cloning specific genes of interest. Hybridization can be used to find a gene of interest. A probe is a molecule complementary in sequence to the gene of interest. The probe, once it has bonded to the gene, can be detected by labeling it with a radioactive isotope or a fluorescent tag.

Genetic engineering can be used for beneficial crop modification, treating genetic disorders, and cloning.

Genetic engineering has benefited agriculture also. For example, many dairy cows are given bovine growth hormone to increase milk production. Commercially grown plants are often genetically modified for optimal growth.

Strains of wheat, cotton, and soybeans have been developed to resist herbicides used to control weeds. This allows for the successful growth of the plants while destroying the weeds. Crop plants are also being engineered to resist infections and pests. Scientists can genetically modify crops to contain a viral gene that does not affect the plant and will "vaccinate" the plant from a virus attack. Crop plants are now being modified to resist insect attacks. This allows for farmers to reduce the amount of pesticide used on plants.

Genetic engineering has made enormous contributions to medicine. The use of DNA probes and polymerase chain reactions (PCR) have enabled scientists to identify and detect elusive pathogens. Diagnosis of genetic disease is now possible before the onset of symptoms.

Genetic engineering has allowed for the treatment of some genetic disorders. **Gene therapy** is the introduction of a normal allele to the somatic cells to replace the defective allele. The medical field has had success in treating patients with a single enzyme deficiency disease. Gene therapy has allowed doctors and scientists to introduce a normal allele that would provide the missing enzyme.

Insulin and mammalian growth hormones have been produced in bacteria by gene-splicing techniques. Insulin treatment helps control diabetes for millions of people who suffer from the disease. The insulin produced in genetically engineered bacteria is chemically identical to that made in the pancreas. Human growth hormone (HGH) has been genetically engineered for treatment of dwarfism caused by insufficient amounts of HGH. HGH is being further researched for treatment of broken bones and severe burns.

Human Genome Project

The goal of the human genome project is to map and sequence the three billion nucleotides in the human genome and to identify all of the genes on it. The project was launched in 1986, and an outline of the genome was finished in 2000, through international collaboration. In May, 2006, the sequence of the last chromosome was published. While the map and sequencing are complete, scientists are still studying the functions of all the genes and their regulation. Humans have successfully decoded the genome of other mammals as well.

It is important to realize that many of the most complex scientific questions have been answered in a collaborative form. The human genome project is a great example of research conducted and shared by multiple countries world wide. It is also interesting to note that because of differing cultural beliefs, some cultures may be more likely to allow areas of research that other cultures may be unlikely to examine.

Skill 3.3   Evolution: evidence, theories and patterns of evolution, factors affecting evolution, speciation, and hypotheses relating to the origin of life

Darwin defined the theory of Natural Selection in the mid-1800s. Through the study of finches on the Galapagos Islands, Darwin theorized that nature selects the traits that are advantageous to the organism. Those genes that do not possess the desirable trait, die and do not pass on their genes. Those more fit to survive will reproduce, thus increasing that gene in the population. Darwin listed four principles to define natural selection:

1. The individuals in a certain species vary from generation to generation.
2. Some of the variations are determined by the genetic makeup of the species.
3. More individuals are produced than will survive.
4. Some genes allow for better survival of an animal.

Darwin, in contrast to other evolutionary scientists, did not believe that traits acquired during an organism's lifetime (e.g. increased musculature) or the desires and needs of the organism affected evolution of populations. For example, Darwin argued that the evolution of long trunks in elephants resulted from environmental conditions that favored those elephants that possessed longer trunks. The individual elephants did not stretch their trunks to reach food or water and pass on the new, longer trunks to their offspring.

Jean Baptiste Lamarck proposed an alternative mechanism of evolution. Lamarck believed individual organisms developed traits in response to changing environmental conditions and passed on these new, favorable traits to their offspring. For example, Lamarck argued that the trunks of individual elephants lengthen as a result of stretching for scarce food and water, and elephants pass on the longer trunks to their offspring. Thus, in contrast to Darwin's relatively random natural selection, Lamarck believed the mechanism of evolution followed a predetermined plan and depended on the desires and needs of individual organisms.

Causes of evolution - Certain factors increase the chances of variability in a population, thus leading to evolution. Items that increase variability include mutations, sexual reproduction, immigration, and large population. Items that decrease variation would be natural selection, emigration, small population, and random mating.

Sexual selection - Genes that come together determine the makeup of the gene pool. Animals that use mating behaviors may be successful or unsuccessful. An animal that lacks attractive plumage or has a weak mating call will not attract the female, thereby eventually limiting that gene in the gene pool. Mechanical isolation, where sex organs do not fit the female, has an obvious disadvantage.

Evidence

The wide range of evidence on evolution provides information on the natural processes by which the variety of life on earth developed.

Paleontology

Paleontology is the study of past life based on fossil records and their relation to different geologic time periods.

When organisms die, they often decompose quickly or are consumed by scavengers, leaving no evidence of their existence. However, occasionally some organisms are preserved; the remains or traces of the organisms from a past geological age embedded in rocks by natural processes are called fossils. They are very important for the understanding the evolutionary history of life on Earth as they provide evidence of evolution and detailed information on the ancestry of organisms.

Petrification is the process by which a dead animal gets fossilized. For this to happen, a dead organism must be buried quickly, to avoid weathering and decomposition. When the organism is buried, the organic matter decays. The mineral salts from the mud (in which the organism is buried) will infiltrate into the bones and gradually fill up the pores. The bones will harden and will then be preserved as fossils. If dead organisms are covered by wind-blown sand, and if the sand is subsequently turned into mud by heavy rain or floods, the same process of mineral infiltration may occur. Besides petrification, the organisms may be well-preserved in ice, in hardened resin of coniferous trees (amber), in tar, in anaerobic acidic peat. Fossilization can sometimes be a trace, an impression of a form (e.g. leaves and footprints).

The horizontal layers of sedimentary rocks (these are formed by silt or mud on top of each other) are called strata, and each layer consists of fossils. The oldest layer is the one at the bottom of the pile. Therefore, fossils found in this layer are the oldest and this is how the paleontologists determine the relative ages of these fossils.

Some organisms appear in a few layers, which only indicate that they lived only during that period and then became extinct. A succession of animals and plants can also be seen in fossil records, which supports the theory that organisms tend to progressively increase in complexity.

According to fossil records, some modern species of plants and animals are found to be almost identical to the species that lived in ancient geological ages. They are existing species of ancient lineage that have remained unchanged morphologically, and may be physiologically unchanged as well. Hence, they are called "living fossils." Some examples of living fossils are tuatara, nautilus, horseshoe crab, gingko and metasequoia.

Anatomy

Comparative anatomical studies reveal that some structural features are basically similar (e.g. flowers generally have sepals, petals, stigma, style and ovary) but the size, color, number of petals, sepals, etc., may differ from species to species.

The degree of resemblance between two organisms indicates how closely they are related in evolution.

- Groups with little in common are supposed to have diverged from a common ancestor much earlier in geological history than groups which have more in common
- To decide how closely related two organisms are, anatomists look for the structures which may serve different purposes in the adult, but are basically similar (homologous)
- In cases where similar structures serve different functions in adults, it is important to trace their origin and embryonic development

A group of organisms sharing a specialized, homologous structure, to perform a variety of functions, in order to adapt to different environmental conditions, is called adaptive radiation. The gradual spreading of organisms with adaptive radiation is known as divergent evolution. Examples of divergent evolution are pentadactyl limb and insect mouthparts

Under similar environmental conditions, fundamentally different structures in different groups of organisms may undergo modifications to serve similar functions. This is called convergent evolution. Analogous structures, which have no close phylogenetic links, show adaptations to perform the same functions. Examples include wings of bats, birds and insects, jointed legs of insects and vertebrates, eyes of vertebrates and cephalopods.

Vestigial organs
Organs that are smaller and simpler in structure than corresponding parts in the ancestral species are called vestigial organs; they are usually degenerated or underdeveloped. These were functional in ancestral species but have become non-functional (e.g. vestigial hind limbs of whales, vestigial leaves of some xerophytes, vestigial wings of flightless birds like ostriches, etc.).

Geographic distribution

- **Continental distribution**: All organisms are adapted to their environment to some extent. It is generally assumed that the same type of species would be found in a similar habitat in a similar geographic area.
  Examples: Africa has short tailed monkeys, lions and giraffes. South America has long-tailed monkeys, jaguars and llamas.
- **Evidence for migration and isolation**: The fossil record shows that the evolution of camels started in North Africa, from which they migrated across the Bering Strait into Asia and Africa and through the Isthmus of Panama into South America.
- **Continental drift**: Fossils of ancient amphibians, arthropods and ferns are found in South America, Africa, India, Australia and Antarctica, which can be dated to the Paleozoic Era, at which time they were all in a single landmass called Gondwana.
- **Oceanic Island distribution**: Most small isolated islands only have native species.

Plant life in Hawaii could have arrived as airborne spores or as seeds in the droppings of birds. The few large mammals present in remote islands were brought by human settlers.

Comparative embryology

Comparative embryology shows how embryos start off looking the same. As they develop, their similarities slowly decrease until they take the form of their particular species.

Example: Adult vertebrates are diverse, yet their embryos are quite similar at very early stages. Fishlike structures still form in early embryos of reptiles, birds and mammals. In fish embryos, a two-chambered heart, some veins, and parts of arteries develop and persist in adult fishes. The same structures form early in human embryos but do not persist as in adults.

Physiology and Biochemistry

Evolution of widely distributed proteins and molecules: All plants and animals make use of DNA and/or RNA. ATP is the metabolic currency. Genetic code is the same for almost every organism. A piece of RNA in a bacterium cell codes for the same protein as in a human cell.

Comparison of the DNA sequence allows organisms to be grouped by sequence and similarity. The resulting phylogenetic trees are typically consistent with traditional taxonomy, and are often used to strengthen or correct taxonomic classifications. DNA sequence comparison is considered strong enough to be used to correct erroneous assumptions in the phylogenetic tree, in cases where other evidence is missing. The sequence of the 168rRNA gene, a vital gene encoding a part of the ribosome was used to find the broad phylogenetic relationships between all life.

The protein and genetic evidence also supports the universal ancestry of life. Vital proteins such as ribosome, DNA polymerase, and RNA polymerase are found in every organism, from the most primitive bacteria to the most complex mammals.

Since metabolic processes do not leave fossils, research into the evolution of the basic cellular processes is done largely by comparing existing organisms.

Speciation

The most commonly used species concept is the **Biological Species Concept (BSC)**. This states that a species is a reproductive community of populations that occupy a specific niche in nature. It focuses on reproductive isolation of populations as the primary criterion for recognition of species status. The biological species concept does not apply to organisms that are asexual in their reproduction, fossil organisms, or distinctive populations that hybridize.

Reproductive isolation is caused by any factor that impedes two species from producing viable, fertile hybrids. Reproductive barriers can be categorized as **prezygotic** (premating) or **postzygotic** (postmating).

The prezygotic barriers are as follows:

1. Habitat isolation – species occupy different habitats in the same territory.
2. Temporal isolation – populations reaching sexual maturity or flowering at different times of the year.
3. Ethological isolation – behavioral differences that reduce or prevent interbreeding between individuals of different species (including pheromones and other attractants).
4. Mechanical isolation – structural differences that make gamete transfer difficult or impossible.
5. Gametic isolation – male and female gametes do not attract each other; no fertilization.

The postzygotic barriers are as follows:

1. Hybrid inviability – hybrids die before sexual maturity.
2. Hybrid sterility – disrupts gamete formation; no normal sex cells.
3. Hybrid breakdown – reduces viability or fertility in progeny of the F_2 backcross.

Geographic isolation can also lead to the origin of a species. **Allopatric speciation** is speciation without geographic overlap. It is the accumulation of genetic differences through division of a species' range, either through a physical barrier separating the population or through expansion by dispersal. In **sympatric speciation**, new species arise within the range of parent populations. Populations are sympatric if their geographical range overlaps. This usually involves the rapid accumulation of genetic differences (usually chromosomal rearrangements) that prevent interbreeding with adjacent populations.

Origin of life

The hypothesis that life developed on Earth from nonliving materials is the most widely accepted theory on the origin of life. The transformation from nonliving materials to life had four stages. The first stage was the nonliving (abiotic) synthesis of small monomers such as amino acids and nucleotides. In the second stage, these monomers combine to form polymers, such as proteins and nucleic acids. The third stage is the accumulation of these polymers into droplets called protobionts. The last stage is the origin of heredity, with RNA as the first genetic material.

The first stage of this theory was hypothesized in the 1920s. A. I. Oparin and J. B. S. Haldane were the first to theorize that the primitive atmosphere was a reducing atmosphere with no oxygen present. The gases were rich in hydrogen, methane, water, and ammonia. In the 1950s, Stanley Miller proved Oparin's theory in the laboratory by combining the above gases. When given an electrical spark, he was able to synthesize simple amino acids. It is commonly accepted that amino acids appeared before DNA. Other laboratory experiments have supported the idea that the other stages in the origin of life theory could have happened.

Other scientists believe that simpler hereditary systems originated before nucleic acids. In 1991, Julius Rebek was able to synthesize a simple organic molecule that replicates itself. According to his theory, this simple molecule may be the precursor of RNA.

Prokaryotes are the simplest life form. Their small genome size limits the number of genes that can control metabolic activities. Over time, some prokaryotic groups became multi-cellular organisms for this reason. Prokaryotes then evolved to form complex bacterial communities where species benefit from one another.

The **endosymbiotic theory** of the origin of eukaryotes states that eukaryotes arose from symbiotic groups of prokaryotic cells. According to this theory, smaller prokaryotes lived within larger prokaryotic cells, eventually evolving into chloroplasts and mitochondria. Chloroplasts are the descendants of photosynthetic prokaryotes and mitochondria are likely to be the descendants of bacteria, which were aerobic heterotrophs. Serial endosymbiosis is a sequence of endosymbiotic events. Serial endosymbiosis may also play a role in the progression of life forms to become eukaryotes.

Skill 3.4 Diversity of life: general characteristics, biological systems of classification, viruses, bacteria, protests, fungi, plants, and animals

Carolus Linnaeus is known as the father of taxonomy. **Taxonomy** is the science of classification. Linnaeus based his system on morphology, the study of structure. Later on, evolutionary relationships (phylogeny) were also used to sort and group species. The modern classification system uses binomial nomenclature. This consists of a two word name for every species. The genus is the first part of the name and the species is the second part. Notice, in the levels explained below, that Homo sapiens is the scientific name for humans. Starting with the kingdom, the groups get smaller and more alike as one moves down the levels in the classification of humans:

Kingdom: Animalia, **Phylum:** Chordata, **Subphylum:** Vertebrata, **Class:** Mammalia, **Order:** Primate, **Family:** Hominidae, **Genus:** Homo, **Species:** sapiens

Species are defined by the ability to successfully reproduce with members of their own kind.

Kingdom Monera - bacteria and blue-green algae, prokaryotic, having no true nucleus, unicellular

Kingdom Protista - eukaryotic, unicellular, some are photosynthetic, some are consumers

Kingdom Fungi - eukaryotic, multicellular, absorptive consumers, contain a chitin cell wall

Bacteria are classified according to their morphology (shape). **Bacilli** are rod shaped, **cocci** are round, and **spirillia** are spiral shaped. The **gram stain** is a staining procedure used to identify bacteria. Gram-positive bacteria pick up the stain and turn purple. Gram-negative bacteria do not pick up the stain and are pink in color. Microbiologists use methods of locomotion, reproduction, and how the organism obtains its food to classify protista.

Methods of locomotion - Flagellates have a flagellum; ciliates have cilia; and ameboids move through use of pseudopodia.

Methods of reproduction - binary fission means that the organism divides in half and is asexual. All new organisms are exact clones of the parent. Sexual modes of reproduction provide more diversity. Bacteria can reproduce sexually through conjugation, where genetic material is exchanged.

Methods of obtaining nutrition - photosynthetic organisms or producers; convert sunlight to chemical energy; consumers, or heterotrophs, eat other living things. Saprophytes are consumers that live off dead or decaying material.

Viruses

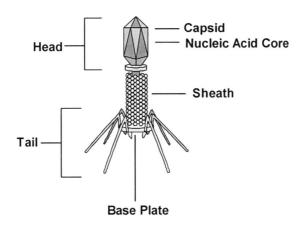

Bacteriophage

All viruses have a head, or protein capsid, that contains genetic material. This material is encoded in the nucleic acid and can be DNA, RNA, or even a limited number of enzymes. Some viruses also have a protein tail. The tail aids in binding to the surface of the host cell and penetrating the surface of the host, in order to introduce the virus's genetic material.

Other examples of viruses and their structures:

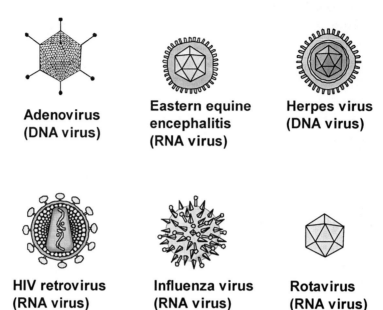

**Adenovirus
(DNA virus)**

**Eastern equine
encephalitis
(RNA virus)**

**Herpes virus
(DNA virus)**

**HIV retrovirus
(RNA virus)**

**Influenza virus
(RNA virus)**

**Rotavirus
(RNA virus)**

Plants

Nonvascular Plants - small in size, do not require vascular tissue (xylem and phloem) because individual cells are close to their environment. The nonvascular plants have no true leaves, stems, or roots.

Division Bryophyta - mosses and liverworts, these plants have a dominant gametophyte generation; they posses rhizoids, which are root-like structures. Moisture in their environment is required for reproduction and absorption.

Vascular Plants - the development of vascular tissue enables these plants to grow in size. Xylem and phloem allows for the transport of water and minerals up to the top of the plant, as well as transporting food manufactured in the leaves to the bottom of the plant. All vascular plants have a dominant sporophyte generation.

Division Lycophyta - club mosses; these plants reproduce with spores and require water for reproduction.

Division Sphenophyta - horsetails; also reproduce with spores. These plants have small, needle-like leaves and rhizoids; they require moisture for reproduction.

Division Pterophyta - ferns; reproduce with spores and flagellated sperm. These plants have a true stem and need moisture for reproduction.

Gymnosperms - The word means "naked seed". These were the first plants to evolve with seeds, which made them less dependent on water to assist in reproduction. Their seeds and pollen from the male can travel by wind. Gymnosperms have cones which protect the seeds.

Division Cycadophyta - cycads; these plants look like palms with cones.

Division Ghetophyta - desert dwellers

Division Coniferophyta - pines; these plants have needles and cones.

Division Ginkgophyta - the Ginkgo is the only member of this division.

Division Anthophyta - Angiosperms are the largest group in the plant kingdom. They are the flowering plants and produce true seeds for reproduction.

Animals

Annelida - segmented worms with specialized tissue. The circulatory system is more advanced in these worms and is a closed system with blood vessels. The nephridia are their excretory organs. They are hermaphrodidic, and each worm fertilizes the other upon mating. They support themselves with a hydrostatic skeleton and have circular and longitudinal muscles for movement.

Mollusca - clams, octopus, soft bodied animals, which have a muscular foot for movement. They breathe through gills and most are able to make a shell for protection from predators. They have an open circulatory system, with sinuses for bathing the body regions.

Arthropoda - insects, crustaceans and spiders; this is the largest group of the animal kingdom. Phylum arthropoda accounts for about 85% of all the animal species. Animals in the phylum arthropoda possess an exoskeleton made of chitin. They must molt in order to grow. Insects, for example, go through four stages of development. They begin as an egg, hatch into a larva, form a pupa, and then emerge as an adult. Arthropods breathe through gills, trachae or book lungs. Movement varies, with members being able to swim, fly, and/or crawl. There is a division of labor among the appendages (legs, antennae, etc). This is an extremely successful phylum, with members occupying diverse habitats.

Echinodermata - sea urchins and starfish. These animals have spiny skin, and live in a marine habitat. They have tube feet for locomotion and feeding.

Chordata - all animals with a notocord or a backbone. The classes in this phylum include Agnatha (jawless fish), Chondrichthyes (cartilage fish), Osteichthyes (bony fish), Amphibia (frogs and toads, which have gills that are replaced by lungs during development), Reptilia (snakes, lizards, which lay eggs with a protective covering), Aves (birds, which are warm-blooded animals with wings consisting of a particular shape and composition designed for flight), and Mammalia (warm-blooded animals with body hair that bear their young alive, and possess mammary glands for milk production).

Skill 3.5	Plants: structure and function of roots, stems, and leaves; nonvascular plants; transport systems, control mechanisms; sexual and a sexual reproduction

Here are some terms related to plants and their structure.

Plant Tissues - specialized tissues enabling plants to grow larger.

Xylem - transports water

Phloem - transports food (glucose)

Cortex - storage of food and water

Epidermis – protective covering

Endodermis - controls movement between the cortex and the cell interior

Pericycle - meristematic tissue, can divide when necessary

Pith - storage in stems

Sclerenchyma and collenchyma - support in stems

Stomata - openings on the underside of leaves, which let carbon dioxide in and water out (transpiration)

Guard cells - control the size of the stomata to control water conservation

Palisade mesophyll - contain chloroplasts in leaves for photosynthesis

Spongy mesophyll - open spaces in the leaf that allows for gas circulation

Seed coat - protective covering on a seed

Cotyledon - small seed leaf that emerges when the seed germinates

Endosperm - food supply in the seed

Apical meristem - this is an area of cell division allowing for growth

Flowers - the reproductive organs of the plant

Pedicel - supports the weight of the flower

Receptacle - holds the floral organs at the base of the flower

Sepals - green leaf-like parts that cover the flower prior to blooming

Petals - contain coloration by pigments, to attract insects for pollination

Anther - male part that produces pollen

Filament - supports the anther; the filament and anther make up the **stamen**

Stigma - female part that holds pollen grains that came from the male part

Style - tube that leads to the ovary (female)

Ovary - contains the ovules; the stigma, style and ovary make up the **carpel**

Photosynthesis is the process by which plants make carbohydrates from the energy of the sun, carbon dioxide, and water. Oxygen is a waste product. Photosynthesis occurs in the chloroplast, where the pigment chlorophyll traps sun energy. It is divided into two major steps:

Light Reactions - Sunlight is trapped, water is split, and oxygen is given off. ATP is made, and hydrogens reduce from NADP to $NADPH_2$. The light reactions occur in light. The products of the light reactions enter into the dark reactions (Calvin cycle).

Dark Reactions - Carbon dioxide enters during the dark reactions which can occur with or without the presence of light. The energy transferred from $NADPH_2$ and ATP allow for the fixation of carbon into glucose.

Respiration - during times of decreased light, plants break down the products of photosynthesis through cellular respiration. Glucose, with the help of oxygen, breaks down and produces carbon dioxide and water as waste. Approximately fifty percent of the products of photosynthesis are used by the plant for energy.

Transpiration - water travels up the xylem of the plant through the process of transpiration. Water sticks to itself (cohesion) and to the walls of the xylem (adhesion). As it evaporates through the stomata of the leaves, the water is pulled up the column from the roots. Environmental factors, such as heat and wind increase the rate of transpiration. High humidity will decrease the rate of transpiration.

Reproduction - Angiosperms are the largest group in the plant kingdom. They are the flowering plants, and they produce true seeds for reproduction. They arose about seventy million years ago, as the dinosaurs were disappearing. The land was drying up, and their ability to produce seeds that could remain dormant until conditions became acceptable allowed for their success. When compared to other plants, they also had more advanced vascular tissue and larger leaves for increased photosynthesis. Angiosperms reproduce through a method of **double fertilization**. An ovum is fertilized by two sperm, one sperm produces the new plant, the other forms the food supply for the developing plant.

Seed dispersal - Success of plant reproduction involves the seed moving away from the parent plant, to decrease competition for space, water, and minerals. Seeds may be carried by wind (maple trees) or water (palm trees), carried by animals (burrs), or ingested by animals and released in their feces in another area.

Skill 3.6   Animals: anatomy and physiological systems, homeostasis, response to stimuli

Skeletal System - The skeletal system functions for support. Vertebrates have an endoskeleton, with muscles attached to bones. Skeletal proportions are controlled by area to volume relationships. Body size and shape is limited due to the forces of gravity. Surface area is increased to improve efficiency in all organ systems.

Muscular System – Its function is for movement. There are three types of muscle tissue. Skeletal muscle is for voluntary motion, these muscles are attached to bones. Smooth muscle is involuntary motion, it is found in organs and enable functions such as digestion and respiration. The cardiac muscle is a specialized type of smooth muscle.

Nervous System - The neuron is the basic unit of the nervous system. It consists of an axon, which carries impulses away from the cell body, the dendrite, which carries impulses toward the cell body and the cell body, which contains the nucleus. Synapses are spaces between neurons. Chemicals called neurotransmitters are found close to the synapse. The myelin sheath, composed of Schwann cells, covers the neurons and provides insulation.

Digestive System - The function of the digestive system is to break down food and absorb it into the blood stream, where it can be delivered to all cells of the body for use in cellular respiration. As animals evolved, digestive systems changed from simple absorption to a system with a separate mouth and anus, allowing the animal to become independent of a host.

Respiratory System - This system functions in the gas exchange of oxygen (needed) and carbon dioxide (waste). It delivers oxygen to the bloodstream and picks up carbon dioxide for release out of the body. Simple animals diffuse gases from and to their environment. Gills allow aquatic animals to exchange gases in a fluid medium by removing dissolved oxygen from the water. Lungs maintain a fluid environment for gas exchange in terrestrial animals.

Circulatory System - The function of the circulatory system is to carry oxygenated blood and nutrients to all cells of the body and return carbon dioxide waste to be expelled from the lungs. Animals evolved from an open system to a closed system with vessels leading to and from the heart.

Animal respiration - takes in oxygen and gives off waste gases. For instance, a fish uses its gills to extract oxygen from the water. Bubbles are evidence that waste gasses are expelled. Respiration without oxygen is called anaerobic respiration. Anaerobic respiration in animal cells is also called lactic acid fermentation. The end product is lactic acid.

Animal reproduction - can be asexual or sexual. Birds and reptiles lay eggs, while animals such as bear cubs, deer, and rabbits are born alive. Some animals reproduce frequently while others do not. Some animals only produce one baby at a time, while others produce many (clutch size).

Animal digestion – some animals only eat meat (carnivores), while others only eat plants (herbivores). Many animals do both (omnivores). Nature has created animals with structural adaptations so they may obtain food through the use of sharp teeth or long facial structures. Digestion's purpose is to break down carbohydrates, fats, and proteins. Many organs are needed to digest food. The process begins with the mouth. Certain animals, such as birds, have beaks to puncture wood or allow for large fish to be consumed. The tooth structure of a beaver is designed to cut down trees. Tigers are known for their sharp teeth, used to rip hides from their prey. Enzymes are catalysts that help speed up chemical reactions by lowering effective activation energy. Enzyme rate is affected by temperature, pH, and the amount of substrate. Saliva is an enzyme that changes starches into sugars.

Animal circulation – The blood temperature of all mammals stays constant, regardless of the outside temperature. This is called warm-blooded, while cold-blooded animals' (amphibians) circulation will vary with the temperature.

Homeostasis

All living organisms respond and adapt to their environments. Homeostasis is the result of regulatory mechanisms that help maintain an organism's internal environment within tolerable limits. For example, in humans and mammals, constriction and dilation of blood vessels near the skin help maintain body temperature.

Response to stimuli

Response to stimuli is one of the key characteristics of any living thing. **Any detectable change in the internal or external environment (the stimulus) may trigger a response in an organism**. Just like physical characteristics, organisms" responses to stimuli are adaptations that allow them to better survive. While these responses may be more noticeable in animals that can move quickly, all organisms are actually capable of responding to changes.

Single-celled organisms

These organisms are able to respond to basic stimuli such as the presence of light, heat, or food. Changes in the environment are typically sensed via **cell surface receptors**. These organisms may respond to such stimuli by making **changes in internal biochemical pathways or initiating reproduction or phagocytosis**. Those capable of **simple motility**, using flagella for instance, may respond by moving toward food or away from heat.

Plants

Plants, typically, do not possess sensory organs, and so, **individual cells recognize stimuli** through a variety of pathways. When **many cells respond to stimuli together**, the response becomes apparent. Logically then, the responses of plants occur on a rather **longer timescale** that those of animals. Plants are capable of **responding to a few basic stimuli including light, water and gravity**. Some common examples include the way plants turn and grow toward the sun, the sprouting of seeds when exposed to warmth and moisture, and the growth of roots in the direction of gravity.

Animals

Lower members of the animal kingdom have responses similar to those seen in single celled organisms. However, higher animals have developed complex systems to detect and respond to stimuli. The **nervous system, sensory organs (eyes, ears, skin, etc), and muscle tissue all allow animals to sense and quickly respond to changes in their environment.**

As in other organisms, many responses to stimuli in animals are **involuntary**. For example, pupils dilate in response to the reduction of light. Such reactions are typically called **reflexes**. However, many animals are also capable of **voluntary response**. In many animal species, voluntary reactions are **instinctual**. For instance, a zebra's response to a lion is a *voluntary* one, but, *instinctually*, it will flee quickly as soon as the lion's presence is sensed. Complex responses, which may or may not be instinctual, are typically termed **behavior**. An example is the annual migration of birds when seasons change. Even more **complex social behavior** is seen in animals that live in large groups.

Skill 3.7 **Ecology: Population dynamics, social behavior, interspecific relationships, community structures and species diversity, succession and disturbance, ecosystems, food webs and energy flow, biomes, biogeochemical cycles**

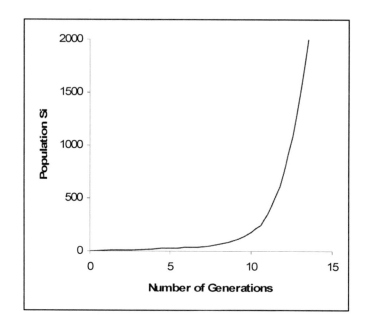

Logistic population growth incorporates the carrying capacity into the growth rate. As a population reaches the carrying capacity, the growth rate begins to slow down and level off.

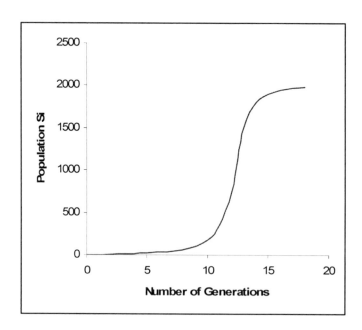

Many populations follow this model of population growth. Humans, however, are an exponentially growing population. Eventually, the carrying capacity of the Earth will be reached, and the growth rate will level off. How and when this will occur remains unknown.

Population density is the number of individuals per unit of area or volume. The spacing pattern of individuals in an area is dispersion. **Dispersion patterns** can be clumped, with individuals grouped in patches; uniformed, where individuals are approximately equidistant from each other; or randomly dispersed.

Population densities are usually estimated based on a few representative plots. Aggregation of a population in a relatively small geographic area can have detrimental effects on the environment. Food, water, and other resources will be rapidly consumed, resulting in an unstable environment. A low population density is less harmful to the environment; the use of natural resources will be more widespread, allowing the environment to recover and continue growth.

Niche

The term 'Niche' describes the relational position of a species or population in an ecosystem. Niche includes how a population responds to the abundance of its resources and enemies (e.g., by growing when resources are abundant and when predators, parasites and pathogens are scarce).

A niche also refers to the life history of an organism, habitat and place in the food chain.

According to the competitive exclusion principle, no two species can occupy the same niche in the same environment for a long time.

The full range of environmental conditions (biological and physical) under which an organism can exist describes its fundamental niche. Because of the pressure from superior competitors, organisms are sometimes driven to occupy a niche much narrower than their previous niche. This is known as the 'realized niche.'

Examples of niche:

1. Oak trees:
   * found in forests
   * absorb sunlight by photosynthesis
   * provide shelter for many animals
   * act as support for creeping plants
   * serve as a source of food for animals
   * cover their ground with dead leaves in the autumn

If the oak trees were cut down or destroyed by fire or storms they would no longer be functional within the niche and this would affect on all the other organisms living in the same habitat.

2. Hedgehogs:
   * eat a variety of insects and other invertebrates, which live underneath the dead leaves and twigs in the garden
   * the spines are a superb environment for fleas and ticks
   * fleas and ticks put the nitrogen back into the soil when they urinate
   * eat slugs and protect plants from them

If there were no hedgehogs around, the population of slugs would explode and the nutrients in the dead leaves and twigs would not be recycled.

Interspecific relationships

Predation and **parasitism** both result in a benefit for one species and a detriment for the other. Predation occurs when a predator eats its prey. The common conception of predation is of a carnivore consuming other animals. This is one form of predation. Herbivory is a form of predation although not always resulting in the death of the plant. Some animals eat enough of a plant to cause the plant to die. Parasitism involves a predator that lives on or in its hosts, causing detrimental effects to the host. Insects and viruses living off and reproducing in their hosts are examples of parasitism. Many plants and animals have defenses against predators. Some plants have poisonous chemicals that will harm the predator if ingested; some animals are camouflaged, so that they are harder to detect by their predators.

Symbiosis occurs when two species live in close relationship to each other. Parasitism is one example of symbiosis described above. Another example of symbiosis is commensalisms. **Commensalism** occurs when one species benefits from the other without harmful effects. **Mutualism** occurs when both species benefit from the other. Species involved in mutualistic relationships must co-evolve to survive. As one species evolves, the other must evolve as well if it is to continue to exist. The grouper and a species of shrimp live in a mutualistic relationship. The shrimp feed off parasites living on the grouper; thus the shrimp are fed and the grouper stays healthy. Many microorganisms have mutualistic relationships with other organisms.

Ecology is the study of organisms, their habitat, and their interactions with the environment. A **population** is a group of the same species in a specific area. A **community** is a group of populations residing in the same area. Communities that are ecologically similar in regards to temperature, rainfall and the species that live there are called **biomes**. Specific biomes include:

Marine - covers 75% of the earth. This biome is organized by the depth of the water. The intertidal zone goes from the tide line to the edge of the water. The littoral zone is from the water's edge to the open sea. It includes coral reef habitats, and it is the most densely populated area of the marine biome. The open sea zone is divided into the epipelagic zone and the pelagic zone. The epipelagic zone receives more sunlight and has a larger number of species. The ocean floor is called the benthic zone and is populated with bottom feeders.

Tropical Rain Forest - temperature is relatively constant (25 degrees C), and rainfall exceeds 200 cm per year. Located around the area of the equator, the rain forest has abundant, diverse species of plants and animals.

Savanna - temperatures range from 0-25 degrees C, depending on the location. Rainfall ranges from 90 to 150 cm per year. Plants include shrubs and grasses. The savanna is a transitional biome between the rain forest and the desert.

Desert - temperatures range from 10-38 degrees C, and rainfall is under 25 cm per year. Plant species include xerophytes and succulents. Lizards, snakes, and small mammals are common animals found in this biome.

Temperate Deciduous Forest - temperature ranges from -24 to 38 degrees C, and rainfall is between 65 to 150 cm per year. Deciduous trees are common, as well as deer, bear, and squirrels.

Taiga - temperatures range from -24 to 22 degrees C, and rainfall is between 35 to 40 cm per year. Taiga is located very north and very south of the equator, getting close to the poles. Plant life includes conifers and plants that can withstand harsh winters. Animals include weasels, mink, and moose.

Tundra - temperatures range from -28 to 15 degrees C, and rainfall is limited, ranging from 10 to 15 cm per year. The tundra is located even further north and south than the taiga. Common plants include lichens and mosses. Animals include polar bears and musk ox.

Polar or Permafrost - temperature ranges from -40 to 0 degrees C. It rarely gets above freezing, and rainfall is below 10 cm per year. Most water is bound up as ice. Life is limited.

Succession - Succession is an orderly process of one community replacing another community that has been damaged or beginning a community where no life previously existed. **Primary succession** occurs after a community has been totally wiped out by a natural disaster or where life never existed before, as in a flooded area. **Secondary succession** takes place in communities that were once flourishing but were disturbed by some influence, either man or nature, but were not totally wiped out. A climax community is a community that is established and flourishing.

Carrying Capacity - this is the total amount of life a habitat can support. Once the habitat runs out of food, water, shelter, or space, the carrying capacity decreases, and then stabilizes.

Ecological Problems – non-renewable resources are fragile and must be conserved for use in the future. Man's impact and knowledge of conservation will directly affect our future.

Biological magnification - chemicals and pesticides accumulate along the food chain. Tertiary consumers have more accumulated toxins than animals at the bottom of the food chain.

Simplification of the food web - three major crops feed the world (rice, corn, and wheat). The planting of these foods wipe out habitats and push animals into other habitats causing overpopulation or extinction.

Fuel sources - strip mining and the overuse of oil reserves have depleted these resources. At the current rate of consumption, conservation or alternate fuel sources will guarantee our future fuel sources.

Pollution - although technology gives us many advances, pollution is a side effect of production. Waste disposal and the burning of fossil fuels have polluted our land, water, and air. Global warming and acid rain are two results of the burning of hydrocarbons and sulfur.

Global warming - rainforest depletion and the use of fossil fuels and aerosols have caused an increase in carbon dioxide production. This leads to a decrease in the amount of oxygen, which is directly proportional to the amount of ozone. As the ozone layer is depleted, more heat enters our atmosphere and is trapped. This causes an overall warming effect, which may eventually melt polar ice caps, causing a rise in water levels and changes in climate, which will affect climates worldwide.

Endangered species - construction of homes to house people in our overpopulated world has caused the destruction of habitat for other animals, leading to their extinction.

Overpopulation - the human race is still growing at an exponential rate. Carrying capacity has not been met due to our ability to use technology to produce more food and housing. Space and water cannot be manufactured, and eventually our non-renewable resources will reach a crisis state. Our over use of nonrenewable resources affects every living thing on this planet.

Biotic factors - living things in an ecosystem: plants, animals, bacteria, fungi, etc. If one population in a community increases, it affects the ability of another population to succeed by limiting the available amount of food, water, shelter, and space.

Abiotic factors - non-living aspects of an ecosystem: soil quality, rainfall, and temperature. Changes in climate and soil can cause effects at the beginning of the food chain, thus limiting or accelerating the growth of populations.

Species diversity

Species diversity is simply a count of the number of different species in a given area. A species is a group of organisms that are similar, able to breed, and produce offspring.

Biologists are not sure of how many different species live on the earth. The estimates range from 2 - 100 million species. So far, only 2.1 million species living in the middle latitudes have been classified; most of the species that are not classified are invertebrates. This group includes insects, spiders, worms, crustaceans, etc. It is difficult to classify them because of their small size and the inaccessible habitats they live in.

In the tropical rain forest, identifying these species is a difficult task since their habitats are hard to explore. Scientists estimate that this single biome may consist 50 - 90% of the Earth's biodiversity.

Many species have gone extinct over Earth's geological history. The main reason for these extinctions is a change in the environment and competition from superior species.

Because of the industrial revolution, a large number of biologically classified species have become extinct. The continued extinction of species on this planet by human activities is one of the greatest environmental problems facing human beings.

Species diversity is one of the three categories of biodiversity. The other two are genetic diversity and ecosystem diversity. Genetic diversity refers to the total number of genetic characteristics expressed and recessed in all of the individuals that comprise a particular species. Ecosystem diversity is the variation of habitats, community types, and abiotic environments present in a given area.

In this context, we need to look at species richness, which is one of the components of the ecosystem's species biodiversity. Species richness is a measurable quality and has been found to be a good substitute for other measures of biodiversity that would be difficult to measure.

A few facts about species richness:

1. The species richness increases from high latitudes to low latitudes.

2. The maximum species richness occurs between 20 and 30 degrees latitude.

3. Species richness increases rapidly from the North pole to the equator, and decreases rapidly from the equator to the South pole.

4. Larger areas contain more species since there are greater opportunities for more species to live there.

5. The relationship between endemism (species that are only native to one habitat) and species richness is positively correlated. However, in some oceanic islands, there is a high degree of endemism, but a low level of species richness.

6. Species richness is a measure of biodiversity.

Food webs and energy flow
Trophic levels are based on the feeding relationships that determine energy flow and chemical cycling.

Autotrophs are the primary producers of the ecosystem. **Producers** mainly consist of plants. **Primary consumers** are the next trophic level. The primary consumers are the herbivores that eat plants or algae. **Secondary consumers** are the carnivores that eat the primary consumers. **Tertiary consumers** eat the secondary consumer. These trophic levels may go higher, depending on the ecosystem. **Decomposers** are consumers that feed off animal waste and dead organisms. This pathway of food transfer is known as the food chain.

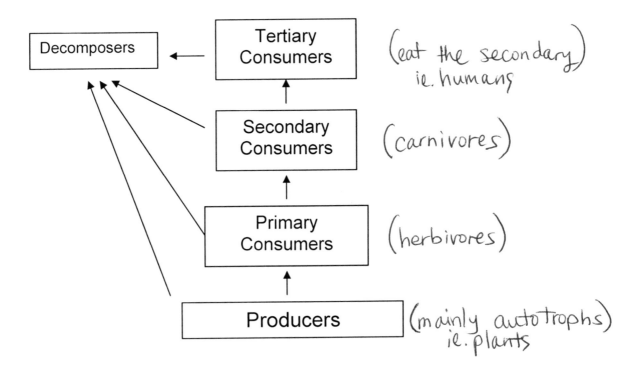

Most food chains are more elaborate and become food webs.

Energy is progressively lost, as the trophic levels progress from producer to tertiary consumer. The amount of energy that is transferred between trophic levels is called the ecological efficiency. The visual of this energy flow is represented in a **pyramid of productivity**.

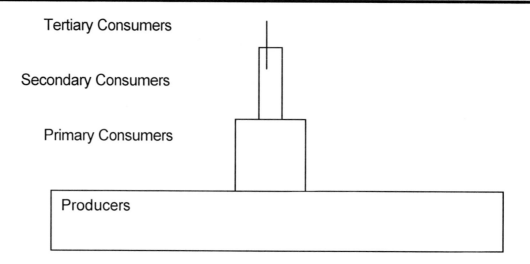

The **biomass pyramid** represents the total dry weight of organisms in each trophic level. A **pyramid of numbers** is a representation of the population size of each trophic level. The producers, being the most populous, are on the bottom of this pyramid with the tertiary consumers on top with the fewest numbers.

Biogeochemical cycles

Biogeochemical cycles are nutrient cycles that involve both biotic and abiotic factors.

Water cycle - 2% of all the available water is fixed and unavailable in ice or the bodies of organisms. Available water includes surface water (lakes, oceans, rivers) and ground water (aquifers, wells) 96% of all available water is from ground water. The water cycle is driven by solar energy. Water is recycled through the processes of evaporation and precipitation. The water present now is the water that has been here since our atmosphere formed.

Carbon cycle - Ten percent of all available carbon in the air (from carbon dioxide gas) is fixed by photosynthesis. Plants fix carbon in the form of glucose; animals eat the plants and are able to obtain their source of carbon. When animals release carbon dioxide through respiration, the plants again have a source of carbon to fix again.

Nitrogen cycle - Eighty percent of the atmosphere is in the form of nitrogen gas. Nitrogen must be fixed and taken out of the gaseous form in order to be incorporated into an organism. Only a few genera of bacteria have the correct enzymes to break the triple bond between nitrogen atoms in a process called nitrogen fixation. These bacteria live within the roots of legumes (peas, beans, alfalfa) and add nitrogen to the soil so it may be taken up by the plant. Nitrogen is necessary to make amino acids and the nitrogenous bases of DNA.

Phosphorus cycle - Phosphorus exists as a mineral and is not found in the atmosphere. Fungi and plant roots have a structure called mycorrhizae that is able to fix insoluble phosphates into useable phosphorus. Urine and decayed matter return phosphorus to the earth where it can be fixed in the plant. Phosphorus is needed for the backbone of DNA and for ATP manufacturing.

In the carbon cycle, decomposers recycle the carbon that accumulates in durable organic material that does not immediately proceed to the carbon cycle. Ammonification is the decomposition of organic nitrogen back to ammonia. This process in the nitrogen cycle is carried out by aerobic and anaerobic bacterial and fungal decomposers. Decomposers add phosphorous back to the soil by decomposing the excretion of animals.

Stability and disturbances

Nature replenishes itself continually. Natural disturbances, such as landslides and brushfires, are not only destructive, but, following the destruction, they allow for a new generation of organisms to inhabit the land. For every indigenous organism, there exists a natural predator. These predator/prey relationships allow populations to maintain reproductive balance and to not over-utilize food sources, thus keeping food chains in check. Left undisturbed, nature would always find a way to balance itself. Unfortunately, the largest disturbance nature faces is from humans. We have introduced non-indigenous species to many areas, upsetting the predator/prey relationships. Our buildings have caused landslides and disrupted waterfront ecosystems. We have damaged the ozone layer and over-utilized the land entrusted to us.

COMPETENCY 4.0   THE EARTH/SPACE SCIENCES

Skill 4.1   **Physical geology: minerals and rocks, folding and faulting, earthquakes and volcanoes, structure of Earth, plate tectonic theory and its supporting evidence, hydrologic cycle, weathering, erosional and depositional processes**

Characteristics by which minerals are classified.

Minerals must adhere to five criteria. They must be (1) inorganic, (2) formed in nature, (3) solid in form, (4) their atoms form a crystalline pattern, (5) its chemical composition is fixed within narrow limits.

There are over 3,000 different types minerals in Earth's crust, which are classified by composition. The major groups of minerals are silicates, carbonates, oxides, sulfides, sulfates, and halides. The largest group of minerals is the silicates. Silicates are made of silicon, oxygen, and one or more other elements.

Rocks are made of minerals. There are three different types of rocks. These are sedimentary, igneous, and metamorphic. Rocks are classified according to where they form, how they form, and their composition.

Sedimentary rocks

Lithification is the process by which sedimentary rocks are formed from sediment. Compaction and cementation of sediment is one of the main ways that sedimentary rocks are formed. Compaction is when the weight of overlying materials compresses and compact the deeper sediments. Often there is some sort of substance that acts as a glue to hold sediment together. This is the process of cementation. Minerals such as quartz or calcite may take on the role of this glue.

Sedimentary rocks are typically formed in layers. Rocks such as shale and limestone are sedimentary rocks. Sandstone, conglomerate, and breccia are sedimentary rocks where you can often see how sediment has been glued together to form the rock.

Sedimentary rocks usually form in the presence of water. Whether it is in the deep ocean, along a river bed, or at a beach, these are the areas where sediment is often glued and compacted together.

Igneous Rocks

Igneous rocks can be classified according to their texture, their composition, and the way they formed. Igneous rocks form from melted rock. This melted rock is called magma when it is under Earth's surface. The same melted rock is called lava when it is on the surface of Earth.

As magma cools, the elements and compounds begin to form crystals. The slower the magma cools, the larger the crystals grow. Rocks with large crystals are said to have a coarse-grained texture. Granite is an example of a coarse grained rock.

When lava is exposed at Earth's surface, it will cool very rapidly. The rocks formed from cooling lava will have a very fine grained texture. The igneous rock basalt is an example of such a fine grained rock. Sometimes the lava will cook so quickly that no crystals have time to form. This produces a rock with a glassy texture, such as obsidian.

Metamorphic rocks

Metamorphic rocks are formed by high temperatures and great pressures. The process by which the rocks undergo these changes is called metamorphism. The outcome of metamorphic changes include: deformation by extreme heat and pressure, compaction, destruction of the original characteristics of the parent rock, bending and folding while in a plastic stage, and the emergence of completely new and different minerals due to chemical reactions with heated water and dissolved minerals.

Metamorphic rocks are classified into two groups, foliated rocks and unfoliated rocks. Foliated rocks consist of compressed, parallel bands of minerals, which give the rocks a striped appearance. Examples of such rocks include slate, schist, and gneiss. Unfoliated rocks are not banded, and examples of such include quartzite, marble, and anthracite rocks.

Orogeny is the process of natural mountain building.

A mountain is terrain that has been raised high above the surrounding landscape by volcanic action or some form of tectonic plate collisions. The plate collisions could be either intercontinental or ocean floor collisions with a continental crust (subduction). The physical composition of mountains would include igneous, metamorphic, or sedimentary rocks; some may have rock layers that are tilted or distorted by plate collision forces.

There are many different types of mountains. The physical attributes of a mountain range depends upon the angle at which plate movement thrust layers of rock to the surface. Many mountains (Adirondacks, Southern Rockies) were formed along high-angle faults.

Folded mountains (Alps, Himalayas) are produced by the folding of rock layers during their formation. The Himalayas are the highest mountains in the world and contain Mount Everest which rises almost 9 km above sea level. The Himalayas were formed when India collided with Asia. The movement which created this collision is still in process at the rate of a few centimeters per year.

Fault-block Mountains (Utah, Arizona, and New Mexico) are created when plate movement produces tension forces instead of compression forces. The area under tension produces normal faults, and rock along these faults is displaced upward.

Dome Mountains are formed as magma tries to push up through the crust but fails to break the surface. Dome Mountains resemble a huge blister on the earth's surface.

Upwarped Mountains (Black Hills of South Dakota) are created in association with a broad arching of the Earth's crust. They can also be formed by rock thrust upward along high angle faults.

Volcanism is the term given to the movement of magma through the crust and its emergence as lava onto the Earth's surface. Volcanic mountains are built up by successive deposits of volcanic materials.

An active volcano is one that is presently erupting or building to an eruption. A dormant volcano is one that is between eruptions but still shows signs of internal activity that might lead to an eruption in the future. An extinct volcano is said to be no longer capable of erupting. Most of the world's active volcanoes are found along the rim of the Pacific Ocean, which is also a major earthquake zone. This curving belt of active faults and volcanoes is often called the Ring of Fire.

The world's best known volcanic mountains include: Mount Etna in Italy and Mount Kilimanjaro in Africa. The Hawaiian Islands are actually the tops of a chain of volcanic mountains that rise from the ocean floor.

There are three types of volcanic mountains: shield volcanoes, cinder cones and composite volcanoes.

Shield Volcanoes are associated with quiet eruptions. Lava emerges from the vent or opening in the crater and flows freely out over the earth's surface until it cools and hardens into a layer of igneous rock. A repeated lava flow builds this type of volcano into the largest type of volcanic mountain. Mauna Loa found in Hawaii, is the largest volcano on earth.

Cinder Cone Volcanoes are associated with explosive eruptions as lava is hurled high into the air in a spray of droplets of various sizes. These droplets cool and harden into cinders and particles of ash before falling to the ground. The ash and cinder pile up around the vent to form a steep, cone-shaped hill called the cinder cone. Cinder cone volcanoes are relatively small but may form quite rapidly.

Composite Volcanoes are described as being built by both lava flows and layers of ash and cinders. Mount Fuji in Japan, Mount St. Helens in Washington, USA and Mount Vesuvius in Italy are all famous composite volcanoes.

Structure of the Earth

The interior of the Earth is divided in to **three chemically distinct layers**. Starting from the middle and moving towards the surface, these are: the core, the mantle, and the crust. Much of what we know about the inner structure of the Earth has been inferred from various data. Subsequently, there is still some uncertainty about the composition and conditions in the Earth's interior.

Core

The **outer core** of the Earth begins about 3,000 km beneath the surface and is a liquid, though far more viscous than that of the mantle. Even deeper, approximately 5,000 km beneath the surface, is the **solid inner core**. The inner core has a radius of about 1,200 km. Temperatures in the core exceed **4,000°C**. Scientists agree that this inner core is **extremely dense**. This conclusion is based on the fact that the Earth is known to have an average density of 5,515 kg/m^3, even though the material close to the surface has an average density of only 3,000 kg/m^3. Therefore, a denser core must exist. Additionally, it is hypothesized that when the Earth was forming, the densest material sank to the middle of the planet. Thus, it is not surprising that the core is about **80% iron**. In fact, there is some speculation that the entire inner core is a single iron crystal, while the outer core is a mix of liquid iron and nickel.

Mantle

The Earth's mantle begins about 35 km beneath the surface and stretches all the way to 3,000 km beneath the surface, where the outer core begins. Since the mantle stretches so far into the Earth's center, its temperature varies widely; near the boundary with the crust, it is approximately 1,000°C, while near the outer core it may reach nearly 4,000°C. Within the mantle, there are silicate rocks that are rich in iron and magnesium. The silicate rocks exist as solids, but the high heat means they are ductile enough to "flow" over a long time period. In general, the mantle is a semi-solid/plastic, and the viscosity varies as pressures and temperatures change at varying depths.

Crust

It is not clear how long the Earth has actually had a solid crust; most of the rocks are less than 100 million years old, though some are 4.4 billion years old. The crust of the Earth is the **outermost layer** and is between 5 and 70 km thick. **Thin areas generally exist under ocean basins (oceanic crust) and thicker crust underlies the continents (continental crust).** Oceanic crust is composed largely of iron magnesium silicate rocks, while continental crust is less dense and consists mainly of sodium potassium aluminum silicate rocks. The crust is the **least dense layer** of the Earth and so is rich in those materials that floated, theoretically, during Earth's formation. Additionally, some heavier elements that bound to lighter materials are present in the crust.

Interactions between the Layers

These layers do not exist as separate entities, with little interaction between them. For instance, it is generally believed that swirling of the iron-rich liquid in the outer core results in the Earth's magnetic field, which is readily apparent on the surface. Heat also moves out from the core to the mantle and crust. The core still retains heat from the formation of the Earth, and additional heat is generated by the decay of radioactive isotopes. While most of the heat in our atmosphere comes from sun, radiant heat from the core does warm oceans and other large bodies of water.

There is also a great deal of interaction between the mantle and the crust. The slow convection of rocks in the mantle is responsible for the shifting of tectonic plates on the crust. Matter can also move between the layers, as during the **rock cycle**. During the rock cycle, igneous rocks are formed when magma, or lava, escapes from the mantle during volcanic eruption. Rocks may also be forced back into the mantle, where the high heat and pressure recreate them as metamorphic rocks.

Mechanisms of producing mountains

Mountains are produced by different types of mountain-building processes. Most major mountain ranges are formed by the processes of folding and faulting.

Folded mountains are produced by the folding of rock layers. Crustal movements press horizontal layers of sedimentary rock together from the sides, squeezing them into wavelike folds. Up-folded sections of rock are called anticlines; down-folded sections of rock are called synclines. The Appalachian Mountains are an example of folded mountains with long ridges and valleys in a series of anticlines and synclines, formed by folded rock layers.

Faults are fractures in the Earth's crust that have been created by either tension or compression forces transmitted through the crust. These forces are produced by the movement of separate blocks of crust.

Faultings are categorized on the basis of the relative movement between the blocks on both sides of the fault plane. This movement can be horizontal, vertical or oblique.

A **dip-slip fault** occurs when the movement of the plates is vertical and opposite. The displacement is in the direction of the inclination, or dip, of the fault. Dip-slip faults are classified as normal faults when the rock above the fault plane moves down, relative to the rock below.

Reverse faults are created when the rock above the fault plane moves up, relative to the rock below. Reverse faults having a very low angle to the horizontal plane are also referred to as thrust faults.

Faults in which the dominant displacement is horizontal movement along the trend or strike (length) of the fault are called **strike-slip faults**. When a large strike-slip fault is associated with plate boundaries it is called a **transform fault**. The San Andreas Fault in California is a well-known transform fault.

Faults that have both vertical and horizontal movement are called **oblique-slip faults**.

When lava cools, igneous rock is formed. This formation can occur either above ground or below ground.

Intrusive rock includes any igneous rock that was formed below the Earth's surface. Batholiths are the largest structures of intrusive rock, and they are composed of near-granite materials. They are the core of the Sierra Nevada Mountains.

Extrusive rock includes any igneous rock that was formed at the Earth's surface.

Dikes are old lava tubes formed when magma entered a vertical fracture and hardened. Sometimes, magma squeezes between two rock layers and hardens into a thin horizontal sheet called a **sill**. A **laccolith** is formed in much the same way as a sill, but the magma that creates a laccolith is very thick and does not flow easily. It pools and forces the overlying strata, creating an obvious surface dome.

A **caldera** is normally formed by the collapse of the top of a volcano. This collapse can be caused by a massive explosion that destroys the cone and empties most, if not all, of the magma in the chamber below the volcano. The cone collapses into the empty magma chamber forming a caldera.

An inactive volcano may have magma solidified in its pipe. This structure, called a volcanic neck, is resistant to erosion, and today, may be the only visible evidence of the past presence of an active volcano.

Glaciation

About 12,000 years ago, a vast sheet of ice covered a large part of the northern United States. This huge, frozen mass had moved southward from the northern regions of Canada as several large glaciers. A time period in which glaciers advance over a large portion of a continent is called an ice age. A glacier is a large mass of ice that moves or flows over the land in response to gravity. Glaciers form among high mountains and in other cold regions

A continental glacier covered a large part of North America during the most recent ice age. Evidence of this glacial coverage remains in many forms: abrasive grooves, large boulders from northern environments dropped in southern locations, glacial troughs created by the rounding out of steep valleys by glacial scouring, and the remains of glacial sources called **cirques**. Cirques were created by frost wedging into the rock at the bottom of the glacier. The remains of plants and animals found in warm climates have been discovered in the moraines and out-wash plains, which helps to support the theory of periods of warmth during the past ice ages.

The Ice Age began about two to three million years ago. This age saw the advancement and retreat of glacial ice over millions of years. Theories relating to the origin of glacial activity include plate tectonics, where it can be demonstrated that some continental masses, now in temperate climates, were at one time blanketed by ice and snow. Other theories involve changes in the Earth's orbit around the sun, changes in the angle of the Earth's axis, and the wobbling of the Earth's axis. Support for the validity of these theories has come from deep-ocean research that indicates a correlation between climatic-sensitive micro-organisms and changes in the Earth's orbital status.

There are two main types of glaciers: valley glaciers and continental glaciers. One characteristic of valley glaciers is U-shaped erosion. These glaciers produce sharp-peaked mountains, such as the Matterhorn in Switzerland. Erosion by continental glaciers often rides over the mountains in their paths, which leaves behind smoothed, rounded mountains and ridges.

Erosion is the inclusion and transportation of surface materials by another moveable material, usually water, wind, or ice. The most important cause of erosion is running water. Streams, rivers, and tides are constantly at work removing weathered fragments of bedrock and carrying them away from their original location.

A stream erodes bedrock by the grinding action of sand, pebbles and other rock fragments. This grinding against each other is called abrasion.

Streams also erode rocks by dissolving or absorbing their minerals. Limestone and marble are readily dissolved by streams.

The breaking down of rocks at or near the Earth's surface is known as **weathering**. Weathering breaks down these rocks into smaller and smaller pieces. There are two types of weathering: physical weathering and chemical weathering.

Physical weathering is the process by which rocks are broken down into smaller fragments without undergoing any change in chemical composition. Physical weathering is mainly caused by the freezing of water, the expansion of rock, and the activities of plants and animals.

Frost wedging is the cycle of daytime thawing and refreezing at night. This cycle causes large rock masses, especially the rocks exposed on mountain tops, to be broken into smaller pieces.

The peeling away of the outer layers from a rock is called **exfoliation**. Rounded mountain tops are called exfoliation domes and have been formed in this way.

Chemical weathering is the breaking down of rocks through changes in their chemical composition. An example would be the change of feldspar in granite to clay. Water, oxygen, and carbon dioxide are the main agents of chemical weathering. When water and carbon dioxide combine chemically, they produce a weak acid that breaks down rocks.

Hydrologic Cycle

Water that falls to Earth in the form of rain and snow is called **precipitation.** Precipitation is part of a continuous process in which water at the Earth's surface evaporates, condenses into clouds, and returns to Earth. This process is called the **water cycle**. The water located below the surface is called groundwater.

The impacts of altitude upon climatic conditions are primarily related to temperature and precipitation. As altitude increases, climatic conditions become increasingly drier and colder. Solar radiation becomes more severe, while the effects of convection forces are minimized. Climatic changes as a function of latitude follow a similar pattern (as a reference, latitude moves either north or south from the equator). The climate becomes colder and drier as the distance from the equator increases. Proximity to land or water masses produces climatic conditions based upon the available moisture. Dry and arid climates prevail where moisture is scarce; while lush tropical climates can prevail where moisture is abundant. Climate, as described above, depends upon the specific combination of conditions making up an area's environment. Man impacts all environments by producing pollutants in Earth, air, and water. It follows then, that man is a major player in world climatic conditions.

Plate Tectonics

Data obtained from many sources led scientists to develop the theory of plate tectonics. This theory is the most current model that explains not only the movement of the continents, but also the changes in the Earth's crust caused by internal forces.

Plates are rigid blocks of the Earth's crust and upper mantle. These rigid, solid blocks make up the lithosphere. The Earth's lithosphere is broken into nine large sections and several small ones. These moving slabs are called plates. The major plates are named after the continents they are "transporting."

The plates float on and move with a layer of hot, plastic-like rock in the upper mantle. Geologists believe that the heat currents circulating within the mantle cause this plastic-like zone of rock to slowly flow, carrying along the overlying crustal plates.

Movement of these crustal plates creates areas where the plates diverge as well as areas where the plates converge. A major area of divergence is located in the mid-Atlantic. Currents of hot mantle rock rise and separate at this point of divergence creating new oceanic crust at the rate of 2 to 10 centimeters per year. Convergence occurs where the oceanic crust collides with either another oceanic plate or a continental plate. The oceanic crust sinks, forming an enormous trench and generating volcanic activity. Convergence also includes continent to continent plate collisions. When two plates slide past one another a transform fault is created.

These movements produce many major features of the Earth's surface, including mountain ranges, volcanoes, and earthquake zones. Most of these features are located at plate boundaries, where the plates interact by spreading apart, pressing together, or sliding past each other. These movements are very slow, averaging only a few centimeters a year.

Boundaries form between spreading plates where the crust is forced apart in a process called rifting. Rifting generally occurs at mid-ocean ridges. Rifting can also take place within a continent, splitting the continent into smaller landmasses that drift away from each other, thereby forming an ocean basin between them. The Red Sea is a product of rifting. As seafloor spreading takes place, new material is added to the inner edges of the separating plates. In this way, the plates grow larger, and the ocean basin widens. This is the process that broke up the super-continent Pangaea and created the Atlantic Ocean.

Boundaries between plates that are colliding are zones of intense crustal activity. When a plate of ocean crust collides with a plate of continental crust, the denser, oceanic plate slides under the lighter continental plate; it then plunges into the mantle. This process is called **subduction**, and the site where it takes place is called a subduction zone. A subduction zone is usually seen on the sea-floor as a deep depression called a trench.

The crustal movement that is identified by plates sliding sideways past one another produces a plate boundary characterized by major faults that are capable of unleashing powerful earthquakes. The San Andreas Fault forms such a boundary between the Pacific Plate and the North American Plate.

Skill 4.2 Historical geology: uniformitarianism, time scales, fossils and stratigraphy, Earth's history

The biological history of the Earth is partitioned into four major eras, which are further divided into major periods. The latter periods are refined into groupings called epochs.

Earth's history extends over more than four billion years and is reckoned in terms of a scale. Paleontologists who study the history of the Earth have divided this huge period of time into four large time units called eons. Eons are divided into smaller units of time called eras. An era refers to a time interval in which particular plants and animals were dominant or present in great abundance. The end of an era is most often characterized by (1) a general uplifting of the crust, (2) the extinction of the dominant plants or animals, and (3) the appearance of new life forms.

Each era is divided into several smaller divisions of time called periods. Some periods are divided into smaller time units called epochs.

Methods of geologic dating

Estimates of the Earth's age have been made possible with the discovery of **radioactivity** and the invention of instruments that can measure the amount of radioactivity in rocks. The use of radioactivity to make accurate determinations of Earth's age is called absolute dating. This process depends upon comparing the amount of radioactive material in a rock with the amount that has decayed into another element. Studying the radiation given off by atoms of radioactive elements is the most accurate method of measuring the Earth's age. These atoms are unstable and are continuously breaking down or undergoing decay. The radioactive element that decays is called the parent element. The new element that results from the radioactive decay of the parent element is called the daughter element.

The time required for one half of a given amount of a radioactive element to decay is called the half-life of that element or compound.

Geologists commonly use carbon dating to calculate the age of a fossil substance.

Infer the history of an area using geologic evidence

The determination of the age of rocks by cataloging their composition has been outmoded since the middle 1800s. Today, a sequential history can be determined by the fossil content (principle of fossil succession) of a rock system, as well as its superposition within a range of systems. This classification process was termed stratigraphy and permitted the construction of a geologic column, in which rock systems are arranged in their correct chronological order.

Principles of catastrophism and uniformitarianism

Uniformitarianism is a fundamental concept in modern geology. It simply states that the physical, chemical, and biological laws that operated in the geologic past operate in the same way today. The forces and processes that we observe presently shaping our planet have been at work for a very long time. This idea is commonly stated as "the present is the key to the past."

Catastrophism is the concept that the Earth was shaped by catastrophic events of a short-term nature.

A fossil is the remains or trace of an ancient organism that has been preserved naturally in the Earth's crust. Sedimentary rocks are usually rich sources of fossil remains. Those fossils, found in layers of sediment, were embedded in the slowly forming sedimentary rock strata. The oldest fossils known are the traces of 3.5 billion-year-old bacteria found in sedimentary rocks. Few fossils are found in metamorphic rock, and virtually none are found in igneous rocks. The magma is so hot that any organism trapped in the magma is destroyed.

The fossil remains of a woolly mammoth embedded in ice were found by a group of Russian explorers. However, the best-preserved animal remains have been discovered in natural tar pits. When an animal accidentally fell into the tar, it became trapped, sinking to the bottom. Preserved bones of the saber-toothed cat have been found in tar pits.

Prehistoric insects have been found trapped in ancient amber or fossil resin that was excreted by some extinct species of pine trees.

Fossil molds are the hollow spaces in a rock previously occupied by bones or shells. A fossil cast is a fossil mold that fills with sediments or minerals that later hardens forming a cast.

Fossil tracks are the imprints in hardened mud left behind by birds or animals.

Skill 4.3   Oceanography: waves, tides, and currents; ocean floor and margins; chemistry of seawater; shore processes; nutrient cycles of the ocean

World weather patterns are greatly influenced by ocean surface currents in the upper layer of the ocean. These currents continuously move along the ocean surface in specific directions. Ocean currents that flow deep below the surface are called sub-surface currents. These currents are influenced by such factors as the location of landmasses in the current's path and the Earth's rotation.

Surface currents are caused by winds and are classified by temperature. Cold currents originate in the Polar regions and flow through surrounding water, which is measurably warmer. The currents with a higher temperature than the surrounding water are called warm currents and can be found near the equator. These currents follow swirling routes around the ocean basins and the equator. The Gulf Stream and the California Current are the two main surface currents that flow along the coastlines of the United States. The Gulf Stream is a warm current in the Atlantic Ocean that carries warm water from the equator to the northern parts of the Atlantic Ocean. Benjamin Franklin studied and named the Gulf Stream. The California Current is a cold current that originates in the Arctic regions and flows southward along the west coast of the United States.

Differences in water density also create ocean currents. Water found near the bottom of the ocean is the coldest and the densest, it tends to flow from a denser area to a less dense area. Currents that flow because of a difference in the density of the ocean's water are called density currents. Water with a higher salinity is denser than water with a lower salinity. Water that has a salinity that is different from the surrounding water may form a density current.

Knowledge of the causes and effects of waves

The movement of ocean water is caused by the wind, the sun's energy, the Earth's rotation, the moon's gravitational pull on Earth, and by underwater earthquakes. Most ocean waves are caused by the impact of winds. When wind blows over the surface of the ocean it transfers energy (friction), to the water and causes waves to form. Waves are also formed by seismic activity on the ocean floor. A wave formed by an earthquake is called a seismic sea wave. These powerful waves can be very destructive, with wave heights increasing to 30 m or more near the shore. The crest of a wave is its highest point, and the trough is its lowest point. The distance from wave top to wave top is the wavelength. The wave period is the time between the passing of two successive waves.

Seafloor

The ocean floor has many of the same features that are found on land. The ocean floor has higher mountains, extensive plains, and deeper canyons than are present on land. Oceanographers have named different parts of the ocean floor according to their structure. The major parts of the ocean floor are:

The **continental shelf** is the sloping part of the continent that is covered with water extending from the shoreline to the continental slope.

The **continental slope** is the steeply-sloping area that connects the continental shelf and the deep-ocean floor.

The **continental rise** is the gently sloping surface at the base of the continental slope.

The **abyssal plains** are the flat, leveled parts of the ocean floor.

A **seamount** is an undersea volcano peak that is at least 1,000 m above the ocean floor.

A **guyot** is a submerged, flat-topped seamount.

Mid-ocean ridges are continuous, undersea mountain chains that are found mostly in the middle portions of the oceans.

Ocean trenches are long, narrow troughs or depressions, formed where ocean floors collide with another section of ocean floor or continent. The deepest trench in the Pacific Ocean is the Marianas Trench, which is about 11 km deep.

Shorelines

The shoreline is the boundary where land and sea meet. Shorelines mark the average position of sea level, which is the average height of the sea without consideration of tides and waves. Shorelines are classified according to the way they were formed. The three types of shorelines are: submerged, emergent, and neutral. When the sea has risen, or the land is sunken, a **submerged shoreline** is created. An **emergent shoreline** occurs when the sea falls or the land rises. A **neutral shoreline** does not show the features of a submerged or an emergent shoreline. A neutral shoreline is usually observed as a flat and broad beach. A **stack** is an island of resistant rock left after weaker rock is worn away by waves and currents. Waves approaching the beach at a slight angle create a current of water that flows parallel to the shore. This long shore current carries loose sediment almost like a river of sand. A **spit** is formed when a weak long shore current drops its load of sand as it turns into a bay.

Rip currents are narrow currents that flow seaward at a right angle to the shoreline. These currents are very dangerous to swimmers.

Most of the beach's sand is composed of grains of resistant material like quartz and orthoclase, although coral or basalt is found in some locations. Many beaches have rock fragments that are too large to be classified as sand.

Seventy percent of the Earth's surface is covered with saltwater, which is termed the hydrosphere. The mass of this saltwater is about 1.4×10^{24} grams. The ocean waters continuously circulate among different parts of the hydrosphere. There are seven major oceans: the North Atlantic Ocean, South Atlantic Ocean, North Pacific Ocean, South Pacific Ocean, Indian Ocean, Arctic Ocean, and the Antarctic Ocean.

Pure water is a combination of hydrogen and oxygen. These two elements make up about 96.5% of ocean water. The remaining portion is made up of dissolved solids. The concentration of these dissolved solids determines the water's salinity.

Salinity is the number of grams of these dissolved salts in 1,000 grams of sea water. The average salinity of ocean water is about 3.5%. In other words, one kilogram of sea water contains about 35 grams of salt. Sodium chloride, or salt (NaCl), is the most abundant of the dissolved salts. The dissolved salts also include smaller quantities of magnesium chloride, magnesium and calcium sulfates, and traces of several other salt elements. Salinity varies throughout the world's oceans; the total salinity of the oceans varies from place to place and also varies with depth. Salinity is low near river mouths where the ocean mixes with fresh water, and salinity is high in areas of high evaporation rates.

The temperature of the ocean water varies with different latitudes and with ocean depths. Ocean water temperature is about constant to depths of 90 meters (m). The temperature of surface water will drop rapidly from 28° C at the equator to - 2° C at the Poles. The freezing point of sea water is lower than the freezing point of pure water. Pure water freezes at 0° C, but the dissolved salts in the sea water keeps it at a freezing point of -2° C. The freezing point of sea water may vary depending on its salinity in a particular location.

The ocean can be divided into three temperature zones. The surface layer consists of relatively warm water and exhibits most of the wave action present. The area where the wind and waves churn and mix the water is called the mixed layer. It is in this layer where most living creatures are found, due to an abundance of sunlight and warmth. The second layer is called the thermocline, where it becomes increasingly cold as its depth increases. This change is due to the lack of energy from sunlight. The layer below the thermocline continues to the deep, dark, very cold, and semi-barren ocean floor.

Oozes - the name given to the sediment that contains at least 30% plant or animal shell fragments. This sediment also contains calcium carbonate. Deposits that form directly from sea water in the place where they are found are called authigenic deposits. Manganese nodules are authigenic deposits found over large areas of the ocean floor.

Causes for the formation of ocean floor features.

The surface of the Earth is in constant motion. This motion is the subject of plate tectonics studies. Major plate separation lines lie along the ocean floors. As these plates separate, molten rock rises, continuously forming new ocean crust and creating new and taller mountain ridges under the ocean. From mapping, it is possible to see that the Mid-Atlantic Range, which divides the Atlantic Ocean basin into two nearly equal parts, shows evidence of these deep-ocean floor changes.

Seamounts are formed by underwater volcanoes. Seamounts and volcanic islands are found in long chains on the ocean floor. They are formed when the movement of an oceanic plate positions a plate section over a stationary hot spot located deep in the mantle. Magma rising from the hot spot punches through the plate and forms a volcano. The Hawaiian Islands are examples of volcanic island chains.

An island arc is magma that rises to produce a curving chain of volcanic islands. An example of an island arc is the Lesser Antilles chain in the Caribbean Sea.

Nutrient Cycles of the ocean

The nutrient cycles in the ocean generate what is known as the biological pump. Interactions between organisms and the environment involve the acquisition of elements required for biological functions. Major nutrients are those elements that are required in high proportions but are not often readily available, because they are bound to themselves or other elements, such as carbon, nitrogen, and phosphorus. Oxygen, hydrogen, and sulfur are also required in high proportions but are readily available in the element form, because of this they do not drive the nutrient cycles.

The **carbon cycle** in the ocean has two main parts, a physical part due to CO_2 dissolving into sea water, and a biological part due to phytoplankton converting CO_2 into carbohydrates. CO_2 dissolves into the ocean at high latitudes, and is then carried to the deep ocean by sinking currents. The CO_2 is reintroduced to the surface ocean by mixing and upwelling. The CO_2 can then be emitted into the tropical atmosphere as needed. The pumping of CO_2 from the atmosphere into the sea for storage is the marine physical pump for carbon.

Phytoplankton in the surface ocean use carbon dioxide, sunlight, water, nitrogen, phosphorus and other nutrients to produce carbohydrates and oxygen. Decay of phytoplankton and the animals that eat them causes a downward rain of organic matter from the surface ocean to the deep ocean. As decay proceeds, the carbon is reduced and oxidized to yield energy (using the phosphorus), water, and carbon dioxide. Most of the carbon on the sea floor is used by animals and returned to the ocean as part of the carbon cycle, thus completing the marine biological pump for carbon.

The **nitrogen cycle** is unique because of its role in the feedback mechanisms of all the cycles. The triple-bonded N_2 molecule, which comprises four fifths of the atmosphere, is very difficult to break apart. Cyanobacteria (known as "nitrogen fixers") expend the energy required to break the bonds. Free nitrogen is needed by plants and phytoplankton to release carbon and oxygen into the atmosphere. However, that process occurs only when O_2 levels are low. When phytoplankton die and rain down to the deep ocean, the nitrogen is converted back to dissolved nitrate in the ocean. The nitrogen cycle is sensitive to the O_2 content of the atmosphere and ocean. Thus, the cycles of carbon, nitrogen, phosphorus and oxygen are linked by feedback mechanisms.

Skill 4.4   Meteorology: structure and properties of the atmosphere; seasonal and latitudinal variation of solar radiation; heat budget; circulation patterns and winds; humidity, clouds, and precipitation; air masses, high and low pressure systems, frontal systems, maps, forecasting; climate and climatic change

Structure and properties of the atmosphere

Dry air is composed of three basic components; dry gas, water vapor, and solid particles (dust from soil, etc.).

The most abundant dry gases in the atmosphere are:

(N_2) Nitrogen   78.09 %
(O_2) Oxygen   20.95 %
(AR) Argon   0.93 %
(CO_2) Carbon Dioxide 0.03 %

The atmosphere is divided into four main layers based on temperature. These layers are labeled troposphere, stratosphere, mesosphere, and thermosphere.

Troposphere - This layer is the closest to the Earth's surface with an average thickness of 7 miles (11 km). All weather phenomena occurs here, as it is the layer with the most water vapor and dust. Air temperature decreases with increasing altitude..

Stratosphere - This layer contains very little water; clouds within this layer are extremely rare. The ozone layer is located in the upper portions of the stratosphere. Air temperature is fairly constant but does increase somewhat with height due to the absorption of solar energy and ultra violet rays from the ozone layer.

Mesosphere - Air temperature again decreases with height in this layer. It is the coldest layer with temperatures in the range of -100^0 C at the top.

Thermosphere - This layer extends upward into space. Oxygen molecules in this layer absorb energy from the sun, causing temperatures to increase with height. The lower part of the thermosphere is called the ionosphere. Here, charged particles, or ions and free electrons can be found. When gases in the Ionosphere are excited by solar radiation, the gases give off light and glow in the sky. These glowing lights are called the Aurora Borealis in the Northern Hemisphere and Aurora Australis in the Southern Hemisphere. The upper portion of the thermosphere is called the exosphere. Gas molecules are very far apart in this layer. Layers of exosphere are also known as the Van Allen Belts and are held together by Earth's magnetic field.

Seasons

The result of the **tilt of the Earth's axis** allows for the seasonable changes called summer, spring, autumn and winter. As the Earth continues to revolve around the sun, it is the angle of the Earth's axis that contributes to the amount of sunlight that is received on Earth, resulting in the changing seasons. **Summer solstice** occurs when the North Pole is tilted toward the sun on June 21st or 22nd, providing increased daylight hours for the Northern Hemisphere and shorter daylight hours for the Southern Hemisphere. **Winter solstice** occurs when the South Pole is tilted toward the sun on December 21st or 22nd, providing shorter daylight hours in the Northern Hemisphere and longer daylight hours in the Southern Hemisphere. The **Spring or Vernal Equinox** occurs on March 20th or 21st, when the direct energy from the sun falls on the equator providing equal lengths of day and night hours in both hemispheres. **The Autumn Equinox** occurs on September 22nd or 23rd providing equal amounts of day and night hours in both hemispheres.

Heat Budget

Heat budget is defined as a relation between fluxes of heat into and out of a given region or body and the heat stored by the system. In general, this budget includes advection, evaporation, radiation, and other forms as well.

To have a clear understanding of heat budget, we need to know the processes associated with it.

1. **Advection**: Advection is the transport of scalar quantity in a vector field. A good example of this is the transport of pollutants or silt in a river. The motion of the water carries these impurities downstream. Any substance can be affected in a similar way in any fluid. Advection is important for the formation of clouds and the precipitation of water from clouds, as part of the hydrological cycle.

In meteorology and physical oceanography, advection often refers to the transport of some property of the atmosphere or ocean, such as heat, humidity, or salinity. Advection is predominantly horizontal.

2. **Evaporation**: Evaporation occurs when a liquid turns into a gas, due to heat energy, allowing the atoms to escape. Evaporation occurs more quickly with increased temperatures. Evaporation is a critical component of the water cycle, which is responsible for clouds and rain.

3. **Thermal Radiation**: Thermal radiation is the electromagnetic radiation emitted from the surface of an object, which is due to the object's temperature. This is an important concept in thermodynamics and it is partially responsible for heat exchange between objects, as warmer bodies radiate more heat than colder bodies. Other factors involved in this process are convection and conduction.

Air Masses

Air masses moving toward or away from the Earth's surface are called air currents. Air moving parallel to Earth's surface is called **wind**. Weather conditions are generated by winds and air currents carrying large amounts of heat and moisture from one part of the atmosphere to another. Wind speeds are measured by instruments called anemometers.

The wind belts in each hemisphere consist of convection cells that encircle the Earth like belts. There are three major wind belts on Earth: (1) trade winds (2) prevailing westerlies, and (3) polar easterlies. Wind belt formation depends on the differences in air pressures that develop in the doldrums, the horse latitudes, and the Polar Regions. The Doldrums surround the equator; within this belt, heated air usually rises straight up into Earth's atmosphere. The Horse latitudes are regions of high barometric pressure with calm and light winds whereas the Polar Regions contain cold, dense air that sinks to the Earth's surface.

Winds caused by local temperature changes include sea breezes and land breezes.

Sea breezes are caused by the unequal heating of the land and an adjacent, large body of water. Since Land heats up faster than water, it creates. the movement of cool, ocean air toward the land which is called a sea breeze. Sea breezes usually begin blowing about mid-morning and end about sunset.

A breeze that blows from the land to the ocean or a large lake is called a **land breeze.**

Monsoons are huge wind systems that cover large geographic areas and that reverse direction seasonally. The monsoons of India and Asia are examples of these seasonal winds, where they alternate wet and dry seasons. As denser, cooler air over the ocean moves inland, a steady, seasonal wind called a summer or wet monsoon is produced.

Cloud types:

Cirrus clouds - white and feathery; high in the sky

Cumulus – thick, white, fluffy

Stratus – layers of clouds cover most of the sky

Nimbus – heavy, dark clouds that represent thunderstorm clouds

Variation of the different cloud types

Cumulo-nimbus – formed from Cumulus clouds; tall dense clouds involved in extreme weather.

Strato-nimbus – dark flat, low clouds, mostly containing liquid droplets.

The air temperature at which water vapor begins to condense is called the **dew point.**

Relative humidity is the actual amount of water vapor in a certain volume of air compared to the maximum amount of water vapor this air could hold at a given temperature.

Types of storms

A **thunderstorm** is a brief, local storm produced by the rapid upward movement of warm, moist air within a cumulo-nimbus cloud. Thunderstorms always produce lightning and thunder and are accompanied by strong wind gusts, heavy rain, and/or hail.

A severe storm with swirling winds that may reach speeds of hundreds of km per hour is called a **tornado**. Such a storm is also referred to as a "twister." The sky is covered by large cumulo-nimbus clouds and violent thunderstorms; a funnel-shaped swirling cloud may extend downward from a cumulo-nimbus cloud and reach the ground. Tornadoes are storms that leave a path of destruction on the ground.

A swirling, funnel-shaped cloud that **extends** downward and touches a body of water is called a **waterspout.**

Hurricanes are storms that develop when warm, moist air, carried by trade winds rotate around a low-pressure "eye". A large, rotating, low-pressure system accompanied by heavy precipitation and strong winds is called a tropical cyclone (better known as a hurricane). In the Pacific region, a hurricane is called a typhoon.

Storms that occur only in the winter are known as blizzards or ice storms. A **blizzard** is a storm with strong winds, blowing snow, and frigid temperatures. An **ice storm** consists of falling rain that freezes when it strikes the ground, covering everything with a layer of ice.

Lines of Latitude

A system of imaginary lines has been developed that helps people describe exact locations on Earth. On a globe, the equator is drawn around Earth halfway between the North and South Poles. Latitude is a term used to describe distance in degrees north or south of the equator. Lines of latitude are drawn east and west, parallel to the equator.

Degrees of latitude range from 0 at the equator to 90 at both the North Pole and South Pole. Lines of latitude are also called parallels.

Lines drawn north and south at right angles to the equator and from pole to pole are called **meridians**. Longitude is a term used to describe distances in degrees east or west of a $0°$ meridian. The prime meridian is the $0°$ meridian which passes through Greenwich, England.

Time zones are determined by longitudinal lines. Each time zone represents one hour. Since there are 24 hours in one complete rotation of the Earth, there are 24 international time zones. Each time zone is roughly $15°$ wide. While time zones are based on meridians, they do not strictly follow lines of longitude. Time zone boundaries are subject to political decisions and in the past have been moved around cities and other areas.

The **International Date Line** is the $180°$ meridian and it is on the opposite side of the world from the prime meridian. The International Date Line is one-half of one day or 12 time zones from the prime meridian. If you were traveling west across the International Date Line, you would lose one day. If you were traveling east across the International Date Line, you would gain one day.

Principles of contouring

A contour line is a line on a map representing an imaginary line on the ground that has the same elevation above sea level along its entire length. Contour intervals usually are given in even numbers or as a multiple of five. In mapping mountains, a large contour interval is used. Small contour intervals may be used where there are small differences in elevation.

Relief describes how much variation in elevation an area has. Rugged, or high relief, describes an area of many hills and valleys. Gentle, or low relief, describes a plain area or a coastal region. Five general rules should be remembered in studying contour lines on a map.

1. Contour lines close around hills and basins or depressions. Hachure lines are used to show depressions. Hachures are short lines placed at right angles to the contour line, which always point toward the lower elevation. A contour line that has hachures is called a depression contour.

2. Contour lines never cross, although they are sometimes very close together. Each contour line represents a certain height above sea level.

3. Contour lines appear on both sides of an area where the slope reverses direction. Contour lines show where an imaginary horizontal plane would slice through a hillside or cut both sides of a valley.

4. Contour lines form V's that point upstream when they cross streams. Streams cut beneath the general elevation of the land surface, and contour lines follow a valley.

5. All contour lines either close (connect) or extend to the edge of the map. No map is large enough to have all its contour lines close.

Interpret maps and imagery

Like photographs, maps readily display information that would be impractical to express in words. Maps that show the shape of the land are called topographic maps. **Topographic maps**, which are also referred to as quadrangles, are generally classified according to publication scale. Relief refers to the difference in elevation between any two points. **Maximum relief** refers to the difference in elevation between the highest and lowest points in the area being considered. **Relief** determines the contour interval, which is the difference in elevation between succeeding contour lines that are used on topographic maps.

Map scales express the relationship between distance (or area) on the map to the true distance (or area) on the Earth's surface. It is expressed as so many feet (miles, meters, km, or degrees) per inch (cm) of map.

Climate

The common weather patterns in a region are called the **climate** of that region. Unlike the weather, which consists of hourly and daily changes in the atmosphere over a region, climate is the average of all weather conditions in a region over a period of time. Many factors are used to determine the climate of a region, including temperature and precipitation. Climate varies from one place to another because of the unequal heating of the Earth's surface. This varied heating of the surface is the result of the unequal distribution of land masses, oceans, and polar ice caps.

Climates are classified into three groups: polar, tropical, and temperate. Climates can be classified further by annual precipitation.

Forecasting weather

Every day, we are affected by weather. It may be in the form of a typical thunderstorm, bringing moist air and cumulonimbus clouds, or a severe storm with pounding winds that can cause either hurricanes or tornados (twisters). These are common terms, as well blizzards or ice storms, which we can all identify with.

The daily newscast relates terms such as dew point, relative humidity, and barometric pressure. Suddenly, common terms become clouded with terms more frequently used by a meteorologist (someone who forecasts weather). **Dew point** is the air temperature at which water vapor begins to condense. **Relative humidity** is the actual amount of water vapor in a certain volume of air, compared to the maximum amount of water vapor that this air could hold at a given temperature.

Weather instruments that forecast weather include the aneroid **barometer** and the mercury barometer, which measure **air pressure**. The air exerts varying pressures on a metal diaphragm that reads air pressure. The mercury barometer operates when atmospheric pressure pushes on a pool of mercury in a glass tube. The higher the pressure, the higher up the tube the mercury will rise.

Relative humidity is measured by two kinds of weather instruments, the **psychrometer** and the hair **gygrometer**. Relative humidity indicates the amount of moisture in the air. Relative humidity is defined as a ratio of existing amounts of water vapor and moisture in the air as compared to the maximum amount of moisture that the air can hold at the same given pressure and temperature. Relative humidity is stated as a percentage, for example the relative humidity can be 100%.

For example, if you were to analyze relative humidity from data, an example might be: if a parcel of air is saturated (meaning it now holds all the moisture it can hold at a given temperature), then the relative humidity is 100%.

Skill 4.5   Astronomy: theories of the origin and structure of the universe, origins and life cycles of stars, major features and structure of the solar system, Sun-Moon-Earth relationships, artificial satellites and space exploration, Earth's seasons, time zones, large units of distance, contributions of remote sensing

Two main hypotheses of the origin of the solar system are: (1) **the tidal hypothesis** and (2) **the condensation hypothesis**.

The **tidal hypothesis** proposes that the solar system began with a near collision of the sun and a large star. Some astronomers believe that as these two stars passed each other, the great gravitational pull of the large star extracted hot gases out of the sun. The mass from the hot gases started to orbit the sun, which began to cool then condensing into the nine planets. However, few astronomers support this example.

The **condensation hypothesis** proposes that the solar system began with rotating clouds of dust and gas. Condensation occurred in the center forming the sun and the smaller parts of the cloud formed the nine planets. This example is widely accepted by many astronomers.

Two main theories to explain the origins of the universe include **The Big Bang Theory** and **The Steady-State Theory.**

The Big Bang Theory has been widely accepted by many astronomers. It states that the universe originated from a magnificent explosion, spreading mass, matter and energy into space. The galaxies formed from this material as it cooled during the next half-billion years.

The Steady-State Theory is the least-accepted theory. It states that the universe is continuously being renewed. Galaxies move outward, and new galaxies replace the older galaxies. Astronomers have not found any evidence to prove this theory.

The future of the universe is hypothesized with the Oscillating Universe Hypothesis. It states that the universe will oscillate or expand and contract. Galaxies will move away from one another and will in time slow down and stop. Then a gradual moving toward each other will again activate the explosion or the Big Bang theory.

The **sun** is the nearest star to Earth, which produces solar energy. By the process of nuclear fusion, hydrogen gas is converted to helium gas. Energy flows out of the core to the surface, allowing radiation to escape into space.

Parts of the sun include: (1) **core:** the inner portion of the sun where fusion takes place, (2) **photosphere:** the surface of the sun which produces **sunspots,** which are cool, dark areas that can be seen on its surface, (3) **chromosphere:** hydrogen gas causes this portion to be red in color. Also found here are solar flares (sudden brightness of the chromosphere) and solar prominences (gases that shoot outward from the chromosphere). (4) **Corona,** the transparent area of sun visible only during a total eclipse.

Solar radiation is energy traveling from the sun that radiates into space. **Solar flares** produce excited protons and electrons that shoot outward from the chromosphere at great speeds reaching earth. These particles disturb radio reception and also affect the magnetic field on earth.

Interrelationships of Sun, Moon, and Earth

The mass of any celestial object may be determined by using Newton's laws of motion and his law of gravity.

For example, to determine the mass of the sun, use the following formula:

$$M = \frac{4\pi^2}{G} = \frac{a^3}{P^2}$$

Where M = the mass of the sun, G = a constant measured in laboratory experiments, a = the distance of a celestial body in orbit around the sun from the sun, and P = the period of the body's orbit.

In our solar system, measurable objects range in mass from the largest, the sun, to the smallest, a near-Earth asteroid. But, this does not take into account, objects with a mass less than 10^{21} kg.

The surface temperature of an object depends largely upon its proximity to the sun. One exception to this, however, is Venus, which is hotter than Mercury because of its cloud layer that holds heat to the planet's surface. The surface temperatures of the planets range from more than 400 degrees on Mercury and Venus to below -200 degrees on the distant planets.

Most minor bodies in the solar system do not have any atmosphere and, therefore, can easily radiate the heat from the sun. In the case of any celestial object, whether a side is warm or cold depends upon whether it faces the sun or not and the time of rotation. The longer the rotation takes, the colder the side facing away from the sun will become and vice versa.

If the density of an object is less than 1.5 grams per cc, then the object is almost exclusively made of frozen water, ammonia, carbon dioxide, or methane. If the density is less than 1.0, the object must be made of mostly gas. In our solar system, there is only one object with such low density -- Saturn. If the density is greater than 3.0 grams per cc, then the object is almost exclusively made of rocks, and if the density exceeds 5.0 grams per cc, then there must be a nickel-iron core. Densities between 1.5 and 3.0 indicate a rocky-ice mixture.

The density of planets correlates with their distance from the sun. The inner planets (Mercury through Mars) are known as the terrestrial planets because they are rocky, and the outer planets (Jupiter and outward) are known as the icy or Jovian (gas-like) planets.

In order for two bodies to interact gravitationally, they must have significant mass. When two bodies in the solar system interact gravitationally, they orbit about a fixed point (the center of mass of the two bodies). This point lies on an imaginary line between the bodies, joining them such that the distances to each body multiplied by each body's mass are equal. The orbits of these bodies will vary slightly over time because of the gravitational interactions.

Satellites and Space exploration

The fundamentals of space knowledge include: survival (especially for prolonged periods of time), an understanding of how spacecrafts get their energy, and how the spacecraft moves through space. Exploration in space involves the history of space, both unmanned and manned missions.

Many financial and ethical questions arise about space travel and the danger involved.

Life support

The health of the astronauts depends on the quality of food they consume. Long space travels must have packaged food that is lightweight, nutritious, will endure temperature and pressure changes, and is easily disposable. The water astronauts consume is filtered from their own breath, urine, and portable water brought on board. The purified water that astronauts use is cleaner than most systems on Earth. The atmosphere is composed of roughly 78% nitrogen, 21% oxygen, and 1% argon and carbon dioxide. In a spacecraft, you would find liquid oxygen and liquid nitrogen; the cabin pressurization systems regulate the use of oxygen and nitrogen. Fire safety is essential to the life support of the astronauts. Fire does not react the same way in space as it does on earth. The operating temperatures are maintained by varied means such as covering the space ship with thermal blankets, paints, and specially made products that reduce both shrinking and expansion.

Energy

There are four major sources of energy a spacecraft will carry. These are batteries, solar panels, RTG's (generators that convert radioisotope waste into electricity, they are heavy and give off radiation), and fuel cells (they convert chemical energy into electricity, are regenerative but heavy).

Propulsion

To better understand propulsion technologies, it is essential to lay a foundation in physics. Topics teachers should review include Newton's laws of motion, rates of change, system of particles, momentum and propulsion, and discussions on impulse and thrust measurements.

Exploration

Unmanned missions are carried out if they are deemed too dangerous for humans to undertake. Sputnik was the first unmanned mission. It was a Russian mission launched on Oct. 4, 1957, during the Cold War. The Americans were driven to compete and enhance their space program.

Mariner 10   This American mission was the first to use the gravitational pull of one planet to reach another planet.

Deep Space 1   This American mission tested twelve advanced technologies to lower the cost and risk of future space travel...

Magellan   The Magellan mission took pictures of and collected information on Venus to help understand the geological structure of that planet....

Mars Exploration Rover   The two Mars Exploration Rovers landed in January of 2004 to robotically explore the geology of Mars. It was also used to explore the possibility of ancient water on Mars.

Mars Pathfinder   The Mars Pathfinder landed July 4, 1997. *The Sojourner* rover onboard was used to analyze the atmosphere, climate, and geology of Mars

Sputnik 1   Sputnik 1 launched October 4, 1957 by Russia and was the first artificial satellite in space. The launch of Sputnik ignited the Space Race between Russia and the United States and strengthened the Cold War.

Sputnik 2   Soviet Union Scientists sent Sputnik 2 into space along with the first live, dog named Laika, to demonstrate that organisms from Earth could survive in orbit.

Voyager 1 / Voyager 2   Voyager 1 and Voyager 2 were spacecrafts designed to explore the outer planets and interstellar space. Launched in 1977, these spacecrafts transmitted information about the gas giants and are currently reaching the edge of our solar system.

Manned missions

Throughout history there have been many manned missions, including many "firsts" such as first animal, man on moon, women in space, preventable catastrophe, and first fatal catastrophe. As of 2007, the manned missions have been contained to orbiting around the earth and landing on the moon. With new knowledge of propulsion, it would be possible to reach Mars with a manned mission. Previously, unmanned missions have used land rovers to collect over 17,000 photo images and rock and soil samples.

Soyuz TM-32	The Russian Soyuz TM-32 was to be kept at the International Space Station as a lifeboat, and the crew that brought it returned to Earth on the Soyuz TM-31 stored there.
Vostok 6	The Russian Vostok 6 was launched to continue experiments for joint spaceflights and also to observe the effect of space travel on the female body.
Apollo-Soyuz	This mission involved a docking of ships between the American Apollo and Russian Soyuz to develop techniques for emergency rescues, as well as to perform some experiments.
Challenger	The American Challenger explosion occurred on the tenth mission of this space shuttle on January 28th 1986.
Vostok 1	This Russian mission was the first manned spaceflight in history, signifying the first time anyone had journeyed into orbit.
Apollo 13	This American mission was to gather information and pictures from the moon. This was the third manned mission to land in outer space. The explosion onboard the space craft was caused by a problem in the oxygen tank..
Apollo 11	This American mission was the first lunar landing that also brought the first man on the moon.
Shenzhou 5	This was the People's Republic of China's first manned flight. Liwei Yang was the first Chinese man in space.
Voskhod 1	In this mission, the USSR launched the first space flight with more than one person aboard. This was also the first flight without spacesuits.

Knowledge of telescope types

Galileo was the first person to use telescopes to observe the solar system. He invented the first refracting telescope. A **refracting telescope** uses lenses to bend light rays in order to focus the image.

Sir Isaac Newton invented the **reflecting telescope** using mirrors to gather light rays on a curved mirror that produces a small focused image.

The world's largest telescope is located in Mauna Kea, Hawaii. It uses multiple mirrors to gather light rays.

The **Hubble space telescope** uses a **single-reflector mirror**. It provides an opportunity for astronomers to observe objects seven times away. Even those objects that are 50 times farther can be viewed better there than by any telescope on Earth. There are future plans to make repairs and install new mirrors and other equipment on the Hubble space telescope.

Refracting and reflecting telescopes are **optical telescopes** since they gather visible light and focus it to produce images. A different type of telescope that collects invisible radio waves created by the sun and stars is called a **radio telescope.**

Radio telescopes consists of a reflector or dish with special receivers. The reflector collects radio waves that are created by the sun and stars. Using a radio telescope has many advantages. They can receive signals 24 hours a day, operates in any kind of weather, and blocks the interference of dust particles and clouds with its performance. The most impressive aspect of the radio telescope is its ability to detect objects from such great distances in space.

The world's largest radio telescope is located in Arecibo, Puerto Rico. It has a collecting dish antenna of more than 300 meters in diameter.

Use spectral analysis to identify or infer features of stars or star systems.

The **spectroscope** is a device or an attachment for telescopes that is used to separate white light into a series of different colors by wave lengths. This series of colors of light is called a **spectrum**. A **spectrograph** can photograph a spectrum, showing wavelengths of light that have distinctive colors. The color red has the longest wavelength and violet has the shortest wavelength. Wavelengths are arranged to form an **electromagnetic spectrum**. They range from very long radio waves to very short gamma rays. Visible light covers a small portion of the electromagnetic spectrum. Spectroscopes observe the spectra, temperatures, pressures, and the movement of stars. The movements of stars indicate if they are moving toward or away from Earth.

If a star is moving towards Earth, light waves compress and the wavelengths of light seem shorter. This will cause the entire spectrum to move towards the blue or violet end of the spectrum.

If a star is moving away from Earth, light waves expand and the wavelengths of light seem longer. This will cause the entire spectrum to move towards the red end of the spectrum.

Astronomical measurement

The three formulas astronomers use for calculating distances in space are the **astronomical unit** (AU), **Light year** (LY), and the **parsec**. It is important to remember that these formulas are measured in distances not time.

The distance between the Earth and the sun is about 150×10^6 km. This distance is known as one astronomical unit or AU. This formula is used to measure distances within the solar system.

The distance light travels in one year is a light year (9.5×10^{12} km). This formula is used to measure distances in space, it does not measure time.

Large distances are measured in parsecs. One parsec equals 3.26 light years.

There are approximately 63,000 AU's in one light year or,

9.5×10^{12} km/ 150×10^6 km = 6.3×10^4 AU

There are eight established planets in our solar system; Mercury, Venus, Earth, Mars, Jupiter, Saturn, Uranus, and Neptune. Pluto was an established planet in our solar system, but as of the summer of 2006, its status was downgraded. The planets are divided into two groups based on distance from the sun. The inner planets include: Mercury, Venus, Earth, and Mars. The outer planets include: Jupiter, Saturn, Uranus, and Neptune.

Planets

Mercury – the closest planet to the sun; its surface has craters and rocks. The atmosphere is composed of hydrogen, helium, and sodium. Mercury was named after the Roman messenger god.

Venus – has a slow rotation when compared to Earth. Venus and Uranus rotate in opposite directions from the other planets, which is called retrograde rotation. The surface of Venus is not visible due to the extensive cloud cover. The atmosphere is composed mostly of carbon dioxide. Sulfuric acid droplets in the dense cloud cover give Venus a yellow appearance. Venus has a greater greenhouse effect than observed on Earth. The dense clouds combined with carbon dioxide trap heat. Venus was named after the Roman goddess of love.

Earth – considered a water planet with 70% of its surface covered by water. Gravity holds the masses of water in place. The different temperatures observed on Earth allow for the different states (solid, liquid, gas) of water to exist. The atmosphere is composed mainly of oxygen and nitrogen. Earth is the only planet that is known to support life.

Mars – the surface of Mars contains numerous craters, active and extinct volcanoes, ridges, and valleys with extremely deep fractures. Iron oxide found in the dusty soil makes the surface seem rust colored and the skies seem pink. The atmosphere is composed of carbon dioxide, nitrogen, argon, oxygen and water vapor. Mars has Polar regions with ice caps composed of water and two moons. Mars was named after the Roman war god.

Jupiter – the largest planet in the solar system.has 16 moons. The atmosphere is composed of hydrogen, helium, methane, and ammonia. There are white colored bands of clouds, which indicates rising gas, and dark colored bands of clouds, which indicates descending gases. The gas movement is caused by heat resulting from the energy of Jupiter's core. Jupiter has a Great Red Spot that is thought to be a hurricane-type cloud. Jupiter has a strong magnetic field.

Saturn – the second largest planet in the solar system, Saturn has rings of ice, rock, and dust particles circling it. Saturn's atmosphere is composed of hydrogen, helium, methane, and ammonia. Saturn has more than 20 moons; it was named after the Roman god of agriculture.

Uranus – the second largest planet in the solar system with retrograde revolution is a gaseous planet. It has 10 dark rings and 15 moons. Its atmosphere is composed of hydrogen, helium, and methane. Uranus was named after the Greek god of the heavens.

Neptune – another gaseous planet with an atmosphere consisting of hydrogen, helium, and methane; it has 3 rings and 2 moons. Neptune was named after the Roman sea god because its atmosphere is the same color as the seas.

Pluto – once considered the smallest planet in the solar system; its status as a planet has been changed to dwarf planet. Pluto's atmosphere probably contains methane, ammonia, and frozen water. Pluto has 1 moon and revolves around the sun every 250 Earth years. Pluto was named after the Roman god of the underworld.

Comets, asteroids, and meteors

Astronomers believe that these rocky fragments may have been the remains of the birth of the solar system that never formed into a planet. **Asteroids** are found in the region between Mars and Jupiter.

Comets are masses of frozen gases, cosmic dust, and small rocky particles. Astronomers think that most comets originate in a dense comet cloud beyond Pluto. A comet consists of a nucleus, a coma, and a tail, which always points away from the sun. The most famous comet, **Halley's Comet,** is named after the person who first discovered it in 240 B.C. It returns to the skies near earth every 75 to 76 years.

Meteoroids are composed of particles of rock and metal of various sizes. When a meteoroid travels through the Earth's atmosphere, friction causes its surface to heat up, and it begins to burn. The burning meteoroid falling through the Earth's atmosphere is called a **meteor** (also known as a "shooting star").

Meteorites are meteors that strike the Earth's surface. A physical example of a meteorite's impact on the Earth's surface can be seen in Arizona. The Barringer Crater is a huge meteorite crater. There are many other meteorite craters throughout the world.
Constellations

Astronomers use groups or patterns of stars, called **constellations,** as reference points to locate other stars in the sky. Familiar constellations include: Ursa Major (meaning big bear) and Ursa Minor (meaning little bear). Within the Ursa Major, the smaller constellation, the Big Dipper is found. Within the Ursa Minor, the smaller constellation, the Little Dipper is found.

Different constellations appear as the Earth continues its revolution around the sun with the seasonal changes.

Magnitude stars are 21 of the brightest stars that can be seen from Earth. These are the first stars noticed at night. In the Northern Hemisphere, there are 15 commonly observed first magnitude stars.

Vast collections of stars are defined as **galaxies**. Galaxies are classified as irregular, elliptical, and spiral. An irregular galaxy has no real structured appearance; most are in their early stages of life. An elliptical galaxy consists of smooth ellipses, containing a little dust and gas, but composed of millions or trillions of stars. Spiral galaxies are disk-shaped and have extending arms that rotate around its dense center. Earth's galaxy is found in the Milky Way, and it is a spiral galaxy.

Terms related to deep space

A **pulsar** is defined as a variable radio source that emits signals in very short, regular bursts; believed to be a rotating neutron star.

A **quasar** is defined as an object that photographs like a star but has an extremely large red shift and a variable energy output; believed to be the active core of a very distant galaxy.

Black holes are defined as an object that has collapsed to such a degree that light can not escape from its surface; that light is trapped by the intense gravitational field.

Remote Sensing

Remote sensing is the measurement or acquisition of information of an object or phenomenon, by a recording device that is not in physical or intimate contact with the object. In reality, remote sensing is the utilization, at a distance (as from aircraft, spacecraft, satellite, or ship), of any device for gathering information about the environment. Common examples of remote sensing are an aircraft taking pictures, Earth observation and weather satellites, monitoring of a fetus in the womb through ultrasound, and space probe.

In the strictest sense, remote sensing could only be applied to terrestrial and weather observations.

There are three techniques used to acquire data.

1. Radiometric
2. Geodetic
3. Acoustic

1. Radiometric: There are a number of methods used for obtaining data for radiometric technique.

* Radar used for ranging and velocity measurements.
* Laser and radar altimeters on satellite for wide range of data collection.
* LIDAR (Light detection and ranging) to measure the concentration of various chemicals in the atmosphere.
* Radiometers and photometers to collect reflected and emitted radiation in a wide range of frequencies.
* Stereographic pairs of aerial photographs (used in making of topographic maps).
* Thematic mappers to map land and physical features.

2. Geodetic: satellite measurements of minute perturbations in the Earth's gravitational field (geodesy) may be used to determine changes in the mass distribution of the Earth, which in turn, may be used for geological or hydrological studies.

3. Acoustic:

* Sonar may be utilized for ranging and measuring underwater objects and terrain.
* Seismograms taken at different locations can locate and measure earthquakes by comparing the relative intensity and precise timing.

Remote sensing works on the principle of inverse problem. While in many cases it may not be possible to measure the desired object, assessment is done indirectly using the surrounding factors.

COMPETENCY 5.0   SCIENCE, TECHNOLOGY, AND SOCIETY

Skill 5.1   Impact of science and technology on the environment and human affairs

Many microorganisms are used to detoxify toxic chemicals and to recycle waste. Sewage treatment plants use microbes to degrade organic compounds. Some compounds, like chlorinated hydrocarbons, cannot be easily degraded. Scientists are working on genetically modifying microbes to be able to degrade the harmful compounds that the current microbes cannot.

Genetic engineering has also benefited agriculture. For example, many dairy cows are given bovine growth hormone to increase milk production. Commercially grown plants are often genetically modified for optimal growth.

Strains of wheat, cotton, and soybeans have been developed to resist herbicides used to control weeds. This allows for the successful growth of the plants while destroying the weeds. Crop plants are also being engineered to resist infections and pests. Scientists can genetically modify crops to contain a viral gene that does not affect the plant but will "vaccinate" the plant from a virus attack. Crop plants are now being modified to resist insect attacks. This allows for farmers to reduce the amount of pesticide used on plants.

Skill 5.2   Human and nature-induced hazards

An important topic in science is the effect of natural disasters on society and the effect human activity has had on inducing such events. Naturally occurring geological, weather, and environmental events can greatly affect people's lives. In addition, the activities of humans can induce such events that would not normally occur.

Nature-induced hazards include floods, landslides, avalanches, volcanic eruptions, wildfires, earthquakes, hurricanes, tornadoes, droughts, and disease. Such events often occur naturally, because of changing weather patterns or geological conditions. Property damage, resource destruction, and the loss of human life are the possible outcomes of natural hazards. Thus, natural hazards are often extremely costly on both an economic and personal level.

While many natural disasters occur naturally, human activity can often stimulate such events. For example, destructive land-use practices, such as mining have induced landslides and avalanches when not properly planned and monitored. In addition, human activities can cause other hazards including global warming and waste contamination.

Global warming is an increase in the Earth's average temperature, resulting, at least in part, from the burning of fuels by humans. Global warming is hazardous because it disrupts the Earth's environmental balance and can negatively affect weather patterns. Ecological and weather pattern changes can promote the natural disasters listed above.

Finally, **improper hazardous-waste disposal** can also contaminate the environment. Hazardous-waste contamination can cause disease in humans. Thus, hazardous waste contamination negatively affects both the environment and the people that live in it.

Skill 5.3   Issues and applications: production, use, management, and disposal of energy and consumer products, management of natural resources

An important application of science and technology is the **production, storage, use, management, and disposal of consumer products and energy**. Scientists from many disciplines work to produce a vast array of consumer products. Energy production and management is another area in which science plays a key role.

The **production** of a large number of popular consumer products requires scientific knowledge and technology. Genetically modified foods, pharmaceuticals, plastics, nylon, cosmetics, household cleaning products, and color additives are but a few examples of science-based consumer goods.

In addition to consumer product production, science helps determine the proper **use and storage of consumer goods**. Safe use and storage is a key component of successful production. For example, perishable products like food must be stored and used in a safe and sanitary way. Science helps establish limits and guidelines for the storage and use of perishable food products.

The **management and disposal** of consumer products is also an important concern. Scientists help establish limits for the safe use of potentially hazardous consumer products. For example, household cleaning products are potentially hazardous if used improperly. Scientific testing determines the proper uses and potential hazards of such products.

Finally, **disposal** of waste from consumer product production and use is of great concern. Proper disposal of hazardous waste and recycling of durable materials is important for the health and safety of human populations, as well as the long-term sustainability of the Earth's resources and environment.

Energy production and management is an increasingly important topic in scientific research because of the increasing scarcity of energy yielding resources, such as petroleum. With traditional sources of energy becoming more scarce and costly, a major goal of scientific energy research is the creation of alternative, efficient means of energy production. Examples of potential sources of alternative energy include wind, water, solar, nuclear, geothermal, and biomass. An important concern in the production and use of energy, both from traditional and alternative sources, is the affect on the environment and the safe disposal of waste products. Scientific research and study helps determine the best method for energy production, use, and waste product disposal, balancing the need for energy with the associated environmental and health concerns.

A **renewable resource** is a resource that is replaced naturally. Living renewable resources would be plants and animals. Plants are renewable because they grow and reproduce. Sometimes renewal of the resource doesn't keep up with the demand. Such is the case with trees. Since the housing industry uses lumber for frames and homebuilding they are often cut down faster than new trees can grow. There are specific tree farms which utilizes special methods that allow trees to grow faster.

A second renewable resource is **animals**. They renew by the process of reproduction. Some wild animals need protection on refuges. As the population of humans increase some resources are used faster. Cattle are used for their hides and for food, while some animals, like deer, are killed for sport. Each state has an environmental protection agency with divisions of forest management and wildlife management.

Non-living renewable resources include water, air, and soil. Water is renewed in a natural cycle called the water cycle. Air is another renewable resource, which is a mixture of gases. Oxygen is given off by plants and taken in by animals that, in turn, expel the carbon dioxide that the plants need. Soil is yet another renewable resource, that when it is fertile is rich in minerals. When plants grow, they remove minerals, which makes the soil less fertile. Chemical treatments are one way of renewing the composition of soil. It is also accomplished naturally when the plants decay back into the soil. The plant material is used to make compost to mix with the soil.

Nonrenewable resources are not easily replaced in a timely fashion. Minerals such as Quartz, mica, salt, and sulfur are nonrenewable resources. Mining depletes these resources so that society may benefit. Glass is made from quartz, electronic equipment from mica, and salt has many uses. Sulfur is used in medicine, fertilizers, paper, and matches.

Metals are among the most widely used nonrenewable resource. Metals must be separated from the ore. Iron is our most important ore. Gold, silver and copper are often found in a pure form called native metals.

Skill 5.4   Social, political, ethical, and economic issues in science and technology

Genetic engineering has drastically advanced biotechnology. Along with these advancements bring safety and ethical questions. Many safety concerns have been answered by government regulations. The FDA, USDA, EPA, and National Institutes of Health are a few of the government agencies that regulate pharmaceutical, food, and environmental technology advancements.

Several ethical questions arise when discussing biotechnology. Should embryonic stem cell research be allowed? Is animal testing humane? These are just a couple of ethical questions that arise when discussing biotechnology. There are strong arguments for both sides of the issues, as well as some government regulations in place to monitor these issues.

Concepts that reflect on a person's ethics may be used for political purposes, either to further one's agenda, or to hurt an opponent. Recent political issues with ethical and scientific ties include abortion, stem-cell research, and cloning. There are at least two sides to each issue, and as such, they can easily become a partisan topic. This partisan position of a topic has the potential to either ensure someone's election or loss.

Local, state, national, and global governments and organizations must increasingly consider policy issues related to science and technology. For example, local and state governments must analyze the impact of proposed development and growth on the environment. Governments and communities must balance the demands of an expanding human population with the local ecology to ensure sustainable growth.

Another segment of society affected by science is the economy. Scientific and technological breakthroughs greatly influence other fields of study and the job market. All academic disciplines utilize computer and information technology to simplify research and information sharing. In addition, advances in science and technology influence the types of available jobs and the desired work skills. For example, machines and computers continue to rapidly replace unskilled laborers. In addition, computer and technological literacy is now a requirement for many jobs and careers. Finally, science and technology continue to change the very nature of careers. Because of science and technology's great influence on all areas of the economy, careers are far less stable than in past eras. Workers can thus expect to change jobs and companies much more often than in the past.

Skill 5.5 Societal issues with health awareness and medical advances

Society faces many issues associated with medical advances. Many medical advances require us to examine our beliefs, and these ethical issues were addressed earlier. Another aspect is the length of human life. As we develop cures for illnesses and learn how to better care for ourselves and the environment, we are pushing back nature's hold. In previous centuries, the average person would live to be sixty, and of a family containing 13 children, only two might live to adulthood. Many died from polio, measles, and mumps. Children today are routinely vaccinated for all three of these diseases. People were also susceptible to death by infection, sometimes from a cut or abscessed tooth. We have created antibiotics to prevent prolonged biotic growth and subsequent illness.

These are common examples, but we have also found ways to cure less common, lethal diseases, and to provide amazing surgical aid, such as organ transplants. While many used to die from cancer, our cure rates are improving. Screening is essential and expensive, but not nearly as expensive as the chemotherapy and radiation drugs available for these diseases. While it is wonderful to live a longer, healthier life, it has created challenges. Although procedures that enable us to live longer are not free.

Sample Test

DIRECTIONS: Read each item and select the best response.

1. **Which is the correct order of the scientific method?**
    1. collecting data
    2. planning a controlled experiment
    3. drawing a conclusion
    4. hypothesizing a result
    5. re-visiting a hypothesis to answer a question
    (Easy) (Skill 1.1)

    A. 1,2,3,4,5
    B. 4,2,1,3,5
    C. 4,5,1,3,2
    D. 1,3,4,5,2

2. **For her first project of the year, a student is designing a science experiment to test the effects of light and water on plant growth. You should recommend that she _____.**
    (Average Rigor) (Skill 1.1)

    A. manipulate the temperature also.
    B. manipulate the water pH also.
    C. determine the relationship between light and water unrelated to plant growth.
    D. omit either water or light as a variable.

3. **When designing a scientific experiment, a student considers all the factors that may influence the results. The goal is to _____.**
    (Average Rigor) (Skill 1.1)

    A. recognize and manipulate independent variables.
    B. recognize and record independent variables.
    C. recognize and manipulate dependent variables.
    D. recognize and record dependent variables.

4. **In an experiment measuring the inhibition of different antibiotics on bacteria grown in Petri dishes, what are the independent and dependent variables respectively?**
    (Rigorous) (Skill 1.1)

    A. Number of bacterial colonies and the antibiotic type.
    B. Antibiotic type and the distance between antibiotic and the closest colony.
    C. Antibiotic type and the number of bacterial colonies.
    D. Presence of bacterial colonies and the antibiotic type.

5. Under a microscope with 440 X magnification an object with diameter 0.1 millimeter appears to have a diameter of _____ .
(Easy) (Skill 1.2)

A. 4.4 millimeters.
B. 44 millimeters.
C. 440 millimeters.
D. 4400 millimeters.

6. When measuring the volume of water in a graduated cylinder, where does one read the measurement?
(Average Rigor) (Skill 1.2)

A. At the highest point of the liquid.
B. At the bottom of the meniscus curve.
C. At the closest mark to the top of the liquid
D. At the top of the plastic safety ring.

7. Which of the following is not an acceptable way for a student to acknowledge sources in a laboratory report?
(Rigorous) (Skill 1.2)

A. The student tells his/her teacher what sources s/he used to write the report.
B. The student uses footnotes in the text, with sources cited, but not in correct MLA format.
C. The student uses endnotes in the text, with sources cited, in correct MLA format.
D. The student attaches a separate bibliography, noting each use of sources.

8. A scientist exposes mice to cigarette smoke, and notes that their lungs develop tumors. Mice that were not exposed to the smoke do not develop as many tumors. Which of the following conclusions may be drawn from these results?:

I. Cigarette smoke causes lung tumors.
II. Cigarette smoke exposure has a positive correlation with lung tumors in mice.
III. Some mice are predisposed to develop lung tumors.
IV. Cigarette smoke exposure has a positive correlation with lung tumors in humans.
(Rigorous) (Skill 1.2)

A. I and II only.
B. II only.
C. I , II, III and IV.
D. II and IV only.

9. Which of these is the best example of 'negligence'? *(Easy) (Skill 1.3)*

A. A teacher fails to give oral instructions to those with reading disabilities.
B. A teacher fails to exercise ordinary care to ensure safety in the classroom.
C. A teacher does not supervise a large group of students.
D. A teacher reasonably anticipates that an event may occur, and plans accordingly.

10. Who should be notified in the case of a serious chemical spill? *(Average Rigor) (Skill 1.3)*

A. The custodian.
B. The fire department or their municipal authority.
C. The science department chair.
D. The School Board.

11. Formaldehyde should not be used in school laboratories for the following reason: *(Average Rigor) (Skill 1.3)*

A. it smells unpleasant.
B. it is a known carcinogen.
C. it is expensive to obtain.
D. it is explosive.

12. **Experiments may be done with any of the following animals except _____ .**
(Rigorous) (Skill 1.3)

A. birds.
B. invertebrates.
C. lower order life.
D. frogs.

13. **In a science experiment, a student needs to repeatedly dispense very small measured amounts of liquid into a well mixed solution. Which of the following is the best choice of equipment for the student?**
(Rigorous) (Skill 1.3)

A. Pipette, Stirring Rod, Beaker.
B. Burette with Burette Stand, Stir-plate, Beaker.
C. Volumetric Flask, Dropper, Stirring Rod.
D. Beaker, Graduated Cylinder, Stir-plate.

14. **A boulder sitting on the edge of a cliff has which type of energy?**
(Easy) (Skill 2.1)

A. Kinetic energy
B. Latent Energy
C. No energy
D. Potential Energy

15. **What is the definition of specific gravity?**
(Average Rigor) (Skill 2.1)

A. The mass of an object.
B. The ratio of the density of a substance to the density of water.
C. Density.
D. The pull of the earth's gravity on an object.

16. **Which of the following statements is most accurate?**
(Average Rigor) (Skill 2.1)

A. Mass is always constant; Weight may vary by location.
B. Mass and Weight are both always constant.
C. Weight is always constant; Mass may vary by location.
D. Mass and Weight may both vary by location.

17. **Which of the following statements describes an isotope of an element?**
(Rigorous) (Skill 2.1)

A. An isotope has a different number of electrons.
B. An isotope has a different number of neutrons.
C. The arrangement of the electrons is different.
D. An isotope has a different number of protons.

18. **Physical properties are observable characteristics of a substance in its natural state. Which of the following are considered physical properties.**
I   Color
II Density
III Specific gravity
IV Melting Point
(Rigorous) (Skill 2.1)

A. I only
B. I and II only
C. I, II, and III only
D. III and IV only

19. **The transfer of heat by electromagnetic waves is called _____ .**
(Easy) (Skill 2.2)

A. conduction.
B. convection.
C. phase change.
D. radiation.

20. **The Law of Conservation of Energy states that**

_____.
(Average Rigor) (Skill 2.2)

A. there must be the same number of products and reactants in any chemical equation.
B. mass and energy can be interchanged.
C. energy is neither created nor destroyed, but may change form.
D. one form energy must remain intact (or conserved) in all reactions

21. **A long silver bar has a temperature of 50 degrees Celsius at one end and 0 degrees Celsius at the other end. The bar will reach thermal equilibrium (barring outside influence) by the process of:-**

_____.
(Average Rigor) (Skill 2.2)

A. heat conduction.
B. heat convection.
C. heat radiation.
D. heat phase change.

22. **When you step out of the shower, the floor feels colder on your feet than the bathmat. Which of the following is the correct explanation for this phenomenon?**
(Rigorous) (Skill 2.2)

A. The floor is colder than the bathmat.
B. The bathmat is smaller than the floor and quickly reaches equilibrium with your body temperature.
C. Heat is conducted more easily to the floor.
D. Water is absorbed from your feet onto the bathmat so it doesn't evaporate as quickly as it does off the floor, thus not cooling the bathmat as quickly.

23. **What is the best explanation of the term "latent heat"?**
(*Rigorous) (Skill 2.2)*

A. The amount of heat it takes to change a solid to a liquid.
B. The amount of heat radiated by an object.
C. The amount heat required for substance to undergo a phase change.
D. The amount of heat it takes to change a liquid to a gas.

24. **Which parts of an atom are located inside the nucleus?**
(*Easy) (Skill 2.3)*

A. electrons and neutrons.
B. protons and neutrons.
C. protons only.
D. neutrons only.

25. **What is the main obstacle to using nuclear fusion to generate electricity?**
(*Average Rigor) (Skill 2.3)*

A. Nuclear fusion produces much more pollution than nuclear fission.
B. There is no obstacle; most power plants us nuclear fusion today.
C. Nuclear fusion requires very high temperature and activation energy.
D. The fuel for nuclear fusion is extremely expensive.

26. **The electrons in an atom that form chemical bonds are called _____.**
(*Average Rigor) (Skill 2.3)*

A. outer shell electrons
B. excited electrons
C. valence electrons.
D. reactive electrons.

27. **In a fission reactor, "heavy water" is used to _____ .**
(*Rigorous) (Skill 2.3)*

A. terminate fission reactions.
B. slow down neutrons and moderate reactions.
C. rehydrate the chemicals.
D. initiate a chain reaction.

28. **All of the following are considered Newton's Laws except:**
(*Easy) (Skill 2.4)*

A. An object in motion will continue in motion unless acted upon by an outside force.
B. For every action force, there is an equal and opposite reaction force.
C. Nature abhors a vacuum.
D. Mass can be considered the ratio of force to acceleration.

29. **All of the following are units of energy except** _____

(Average Rigor) (Skill 2.4)

A. joules.
B. calories.
C. watts.
D. ergs.

30. **The force of gravity on earth causes all bodies in free fall to _____ .**
(Average Rigor) (Skill 2.4)

A. fall at the same speed.
B. accelerate at the same rate.
C. reach the same terminal velocity.
D. move in the same direction.

31. **Newton's Laws are taught in science classes because _____ .**

(Rigorous) (Skill 2.4)

A. they are the correct analysis of inertia, gravity, and forces.
B. they are a close approximation to correct physics, for usual Earth conditions.
C. they accurately incorporate relativity into the study of force.
D. Newton was a well-respected scientist in his time.

32. **Since ancient times, people have been enthralled with bird flight. What is the key to bird flight?**
(Rigorous) (Skill 2.4)

A. Birds' wings are a particular shape and composition that causes the air flow over the wing to travel faster than the air flow under the wing.
B. Birds flap their wings quickly enough to propel themselves.
C. Birds' wings are a particular shape and composition that causes the air flow under the wing to travel faster than the air flow over the wing.
D. Birds flap their wings to create a downward force that opposes gravity.

33. **Resistance is measured in units called _____ .**
(Average Rigor) (Skill 2.5)

A. watts.
B. volts.
C. ohms.
D. current.

34. **The electromagnetic radiation with the longest wave length is _____ .**

(Average Rigor) (Skill 2.5)

A. radio waves.
B. red light.
C. X-rays.
D. ultraviolet light.

35. A 10 ohm resistor and a 50 ohm resistor are connected in parallel. If the current in the 10 ohm resistor is 5 amperes, the current (in amperes) running through the 50 ohm resistor is
(Rigorous) (Skill 2.5)

A. 1
B. 50
C. 25
D. 60

36. A light bulb is connected in series with a rotating coil within a magnetic field. The brightness of the light may be increased by any of the following except:
(Rigorous) (Skill 2.5)

A. Rotating the coil more rapidly.
B. Adding more loops to the coil.
C. Using tighter loops for the coil.
D. Using a stronger magnetic field.

37. Sound waves are produced by _____ .
(Easy) (Skill 2.6)

A. pitch.
B. noise.
C. vibrations.
D. sonar.

38. The Doppler Effect is most closely with which property of waves?
(Average Rigor) (Skill 2.6)

A. amplitude.
B. wavelength.
C. frequency.
D. intensity.

39. The speed of light changes in different materials. This is due to _____ .
(Average Rigor) (Skill 2.6)

A. interference.
B. refraction.
C. reflection.
D. relativity.

40. A converging lens produces a real image _____.
(Rigorous) (Skill 2.6)

A. never
B. when the object is exactly at a distance of one focal length
C. when the object is within one focal length of the lens.
D. when the object is further than one focal length from the lens.

41. As a train approaches and blasts its whistle, the whistle sounds _____ .
(Rigorous) (Skill 2.6)

   A. higher, because it has a higher apparent frequency.
   B. lower, because it has a lower apparent frequency.
   C. higher, because it has a lower apparent frequency.
   D. lower, because it has a higher apparent frequency.

42. The elements in the modern Periodic Table are arranged _____ .
(Easy) (Skill 2.7)

   A. in numerical order by atomic number.
   B. randomly.
   C. in alphabetical order by chemical symbol.
   D. in the order of their discovery.

43. Which group of metals is the most chemically active?
(Average Rigor) (Skill 2.7)

   A. Alkaline Earth Metals
   B. Transition elements
   C. Alkali Metals
   D. Metalloids

44. Which of the following is not a property of metalloids?
(Rigorous) (Skill 2.7)

   A. Metalloids are solids at standard temperature and pressure.
   B. Metalloids can conduct electricity to a limited extent.
   C. Metalloids are found in groups 13 through 17.
   D. Metalloids all favor ionic bonding.

45. Which of the following statements are true of all transition elements?
(Rigorous) (Skill 2.7)

   A. They are all hard solids at room temperature.
   B. They tend to form salts when reacted with Halogens.
   C. They all have a silvery appearance in their pure state.
   D. All of the Above

46. **What is the best definition of Isomerization?**
(Easy) (Skill 2.8)

   A. A chemical reaction in which a molecule changes shape, but no atoms are lost or gained.
   B. A chemical reaction in which a molecule changes shape, but one or more atoms are lost.
   C. A chemical reaction in which a molecule changes shape, but one or more atoms are gained.
   D. A chemical reaction in which a compound is broken down into its constitute parts.

47. **Which of the following is found in the least abundance in organic molecules?**
(Average Rigor) (Skill 2.8)

   A. Phosphorus.
   B. Potassium.
   C. Argon.
   D. Oxygen.

48. **What is necessary for ion diffusion to occur spontaneously?**
(Average Rigor) (Skill 2.8)

   A. Carrier proteins.
   B. Energy from an outside source.
   C. A concentration gradient.
   D. Activation Energy

49. **The chemical equation for water formation is: 2H2 + O_2 → $2H_2O$. Which of the following is an incorrect interpretation of this equation?**
(Rigorous) (Skill 2.8)

   A. Two moles of hydrogen gas and one mole of oxygen gas combine to make two moles of water.
   B. Two grams of hydrogen gas and one gram of oxygen gas combine to make two grams of water.
   C. Two molecules of hydrogen gas and one molecule of oxygen gas combine to make two molecules of water.
   D. Four atoms of hydrogen (combined as a diatomic gas) and two atoms of oxygen (combined as a diatomic gas) combine to make two molecules of water.

50. **Carbon forms bonds with hydrogen by _____ .**
(Rigorous) (Skill 2.8)

   A. ionic bonding.
   B. non-polar covalent bonding.
   C. polar covalent bonding.
   D. strong nuclear force.

51. A cup of hot liquid and a cup of cold liquid are both sitting in a room at a temperature of 72 degrees Fahrenheit and 25% humidity. Both cups are made of thin plastic. Which of the following is a true statement?
(Easy) (Skill 2.9)

A. There will be condensation on the outside of both cups.
B. There will be condensation on the outside of the hot liquid cup, but not on the cold liquid cup.
C. There will be condensation on the outside of the cold liquid cup, but not on the hot liquid cup.
D. There will not be condensation on the outside of either cup.

52. When heat is added to most solids, they expand. Why is this the case?
(Average Rigor) (Skill 2.9)

A. The molecules become larger.
B. The increased molecular motion leads to greater distance between the molecules.
C. The molecules develop greater repelling electric forces when heated.
D. The molecules form a more rigid structure.

53. Which of the following is not true about phase change in matter?
(Rigorous) (Skill 2.9)

A. Solid water and liquid ice can coexist at water's freezing point.
B. At 7 degrees Celsius, water is always in liquid phase.
C. Matter changes phase when enough energy is gained or lost.
D. Different phases of matter are characterized by differences in molecular motion.

54. If the volume of a confined gas is increased, what happens to the pressure of the gas? You may assume that the gas behaves ideally, and that temperature and number of gas molecules remain constant.
(Rigorous) (Skill 2.9)

A. The pressure increases.
B. The pressure decreases.
C. The pressure stays the same.
D. There is not enough information given to answer this question.

55. Catalysts assist reactions by _____ .
    (Easy) (Skill 2.10)

    A. lowering required activation energy.
    B. maintaining precise pH levels.
    C. keeping systems at equilibrium.
    D. changing the starting amounts of reactants.

56. Which of the following will not change in a chemical reaction?
    (Average Rigor) (Skill 2.10)

    A. Number of moles of products.
    B. Atomic number of one of the reactants.
    C. Mass (in grams) of one of the reactants.
    D. Temperature of the system.

57. Which of the following is a correct definition for 'chemical equilibrium'?
    (Average Rigor) (Skill 2.10)

    A. Chemical equilibrium occurs when the forward and backward reaction rates are equal. The reaction may continue to proceed forward and backward.
    B. Chemical equilibrium occurs when the forward and backward reaction rates are equal, and equal to zero. The reaction does not continue.
    C. Chemical equilibrium occurs when there are equal quantities of reactants and products.
    D. Chemical equilibrium occurs when acids and bases neutralize each other fully.

58. Which change does not affect enzymatic rates?
    (Rigorous) (Skill 2.10)

    A. Increase the temperature.
    B. Add more substrate.
    C. Adjust the pH.
    D. Use a larger cell.

59. Which reaction below is a decomposition reaction?
    (Rigorous) (Skill 2.10)

    A. $HCl + NaOH \rightarrow NaCl + H_2O$
    B. $C + O_2 \rightarrow CO_2$
    C. $2H_2O \rightarrow 2H_2 + O_2$
    D. $CuSO_4 + Fe \rightarrow FeSO_4 + Cu$

60. Vinegar is an example of a
_____ .
(Easy) (Skill 2.11)

A. strong acid.
B. strong base.
C. weak acid.
D. weak base.

61. Which of the following statements are true of vapor pressure at equilibrium?
(Average Rigor) (Skill 2.11)

A. Solids have no vapor pressure.
B. Dissolving a solute in a liquid increases its vapor pressure.
C. The vapor pressure of a pure substance is characteristic of that substance and its temperature.
D. The vapor pressure of a gas is unique to each gas and is independent of temperature.

62. Which one of the following compounds would form the strongest electrolytes when dissolved in water?
(Average Rigor) (Skill 2.11)

A. Glucose
B. Lemon juice
C. NaBr
D. Heptane

63. Which of the following occur when NaCl dissolves in water?
(Rigorous) (Skill 2.11)

A. Heat is required to break bonds in the NaCl crystal lattice.
B. Heat is released when hydrogen bonds in water are broken.
C. Heat is required to form bonds of hydration.
D. The oxygen end of the water molecule is attracted to the Cl^- ion.

64. The first stage of mitosis is called _____ .
(Average Rigor) (Skill 3.1)

A. telophase.
B. anaphase.
C. prophase.
D. metaphase.

65. Which process(es) result(s) in a haploid chromosome number?
(Average Rigor) (Skill 3.1)

A. Mitosis.
B. Meiosis.
C. Both mitosis and meiosis.
D. Neither mitosis nor meiosis.

66. A by-product of anaerobic respiration in animals is
_____.
(Rigorous) (Skill 3.1)

A. carbon dioxide.
B. lactic acid.
C. oxygen.
D. sodium chloride

67. Which cellular organelle contains the food and other materials needed by the cell.
(Rigorous) (Skill 3.1)

A. Vacuoles.
B. Golgi Apparatus.
C. Ribosomes.
D. Lysosomes.

68. Which of the following is not a nucleotide?
(Average Rigor) (Skill 3.2)

A. adenine.
B. alanine.
C. cytosine.
D. guanine.

69. A white flower is crossed with a red flower. Which of the following is a sign of incomplete dominance?
(Average Rigor) (Skill 3.2)

A. Pink flowers.
B. Red flowers.
C. White flowers.
D. No flowers.

70. A child has type O+ blood. Her father has type A+ blood, and her mother has type B- blood. What are the genotypes of the father and mother, respectively?
(Rigorous) (Skill 3.2)

A. AO+ - and BO - -.
B. AO+ + and BO - -.
C. AO+ + and BO + -.
D. Cannot determine both parents genotype from the information provided

71. Amino acids are carried to the ribosome in protein synthesis by _____ .
(Rigorous) (Skill 3.2)

A. transfer RNA (tRNA).
B. transport enzymes.
C. ribosomal RNA (rRNA).
D. cytoskeletal transport proteins.

72. A duck's webbed feet are an example of _____ .
(Easy) (Skill 3.3)

A. mimicry.
B. structural adaptation.
C. protective resemblance.
D. protective coloration.

73. **Which of the following is not one of the principles of Darwin's Theory of Natural Selection?**
(Average Rigor) (Skill 3.3)

A. More individuals are produced than will survive.
B. The individuals in a certain species vary from generation to generation.
C. Only the fittest members of a species survive.
D. Some genes allow for better survival of an animal.

74. **Which of the following is the best example of an explanation of the theory of evolution?**
(Rigorous) (Skill 3.3)

A. Giraffes need to reach higher for leaves to eat, so their necks stretch. The giraffe babies are then born with longer necks. Eventually, there are more long-necked giraffes in the population.
B. Giraffes with longer necks are better able to reach more leaves, so they eat more and have more babies than other giraffes. Eventually, there are more long-necked giraffes in the population.
C. Giraffes want to reach higher for leaves to eat, so they release enzymes into their bloodstream, which in turn causes fetal development of longer-necked giraffes. Eventually, there are more long-necked giraffes in the population.
D. Giraffes with long necks are more attractive to other giraffes, so they get the best mating partners and have more babies. Eventually, there are more long-necked giraffes in the population.

75. **Members of the same animal species _____ .**
(Easy) (Skill 3.4)

A. look identical.
B. never adapt differently.
C. are able to reproduce with each other.
D. are found in the same geographic location.

76. **Which of the following is not a necessary characteristic of living things?**
(Average Rigor) (Skill 3.4)

A. Movement.
B. Reduction of local entropy.
C. Composed of cells.
D. Reproduction.

77. **Animals with a notochord or backbone are in the phylum**
(Average Rigor) (Skill 3.4)

A. Arthropoda.
B. Chordata.
C. Mollusca.
D. Ammalia.

78. **Laboratory researchers have classified fungi as distinct from plants because the cell walls of fungi _____ .**
(Rigorous) (Skill 3.4)

A. contain chitin.
B. contain yeast.
C. are more solid.
D. are less solid.

79. **Which part of a plant is responsible for transporting water?**
(Easy) (Skill 3.5)

A. Phloem
B. Xylem
C. Stomata
D. Cortex

80. **Which of the following organisms uses spores to reproduce?**
(Average Rigor) (Skill 3.5)

A. Fish.
B. Flowering plants.
C. Conifers.
D. Ferns.

81. **Which of the following is not characteristic of Gymnosperms?**
(Rigorous) (Skill 3.5)

A. They are less dependent on water for reproduction than other plant groups.
B. Gymnosperms have cones which protect their seeds.
C. Gymnosperms reproduce asexually.
D. Gymnosperm seeds and pollen are easily carried by the wind.

82. **Which is the correct sequence of insect development?**
(Easy) (Skill 3.6)

A. Egg, pupa, larva, adult.
B. Egg, larva, pupa, adult.
C. Egg, adult, larva, pupa.
D. Pupa, egg, larva, adult.

83. **Echinodermata are best known for what characteristic?**
(Average Rigor) (Skill 3.6)

A. Their slimy skin
B. Their Dry Habitat
C. Their tube feet
D. Their tentacles.

84. **Mollusca have an open circulatory system. Their sinuses serve which purpose?**
(Rigorous) (Skill 3.6)

A. Breathing
B. Bathing
C. Filtering food
D. Circulating blood

85. **What are the most significant and prevalent elements in the biosphere?**
(Easy) (Skill 3.7)

A. Carbon, Hydrogen, Oxygen, Nitrogen, Phosphorus.
B. Carbon, Hydrogen, Sodium, Iron, Calcium.
C. Carbon, Oxygen, Sulfur, Manganese, Iron.
D. Carbon, Hydrogen, Oxygen, Nickel, Sodium, Nitrogen.

86. **A wrasse (fish) cleans the teeth of other fish by eating away plaque. This is an example of _____ between the fish.**
(Average Rigor) (Skill 3.7)

A. parasitism.
B. symbiosis (mutualism).
C. competition.
D. predation.

87. **What is the most accurate description of the Water Cycle?**
(Rigorous) (Skill 3.7)

A. Rain comes from clouds, filling the ocean. The water then evaporates and becomes clouds again.
B. Water circulates from rivers into groundwater and back, while water vapor circulates in the atmosphere.
C. Water is conserved except for chemical or nuclear reactions, and any drop of water could circulate through clouds, rain, groundwater, and surface-water.
D. Water flows toward the oceans, where it evaporates and forms clouds, which causes rain, which in turn flow back to the oceans after it falls.

88. **What is the source of drinking water for most of the United States?**
(Rigorous) (Skill 3.7)

A. Desalinated ocean water.
B. Surface water (lakes, streams, mountain runoff).
C. Rainfall into municipal reservoirs.
D. Groundwater.

89. **_____ are areas of weakness in the plates of the earth's crust.**
(Easy) (Skill 4.1)

A. Faults.
B. Ridges.
C. Earthquakes.
D. Volcanoes.

90. **Which of the following is not a type of volcano?**
(Average Rigor) (Skill 4.1)

A. Shield volcanoes.
B. Composite volcanoes.
C. Stratus volcanoes.
D. Cinder cone volcanoes.

91. **Which of these is a true statement about loamy soil?**
(Average Rigor) (Skill 4.1)

A. Loamy soil is gritty and porous.
B. Loamy soil is smooth and a good barrier to water.
C. Loamy soil is hostile to microorganisms.
D. Loamy soil is velvety and clumpy.

92. **Lithification refers to the process that creates**

_____.
(Rigorous) (Skill 4.1)

A. metamorphic rocks.
B. sedimentary rocks.
C. igneous rocks.
D. lithium oxide.

93. **Which of the following is an example of an Igneous rock?**
(Rigorous) (Skill 4.1)

A. Quartz.
B. Shale
C. Gneiss
D. Obsidian

94. **Fossils are usually found in _____ rock.**
(Easy) (Skill 4.2)

A. igneous.
B. sedimentary.
C. metamorphic.
D. large grained.

95. **The end of a geologic era is most often characterized by _____**
(Average Rigor) (Skill 4.2)

A. a general uplifting of the crust.
B. the extinction of the dominant plants and animals
C. the appearance of new life forms.
D. all of the above.

96. **Which of the following is the longest (largest) unit of geological time?**
(Average Rigor) (Skill 4.2)

A. Era.
B. Epoch.
C. Period.
D. Eon.

97. **Which of the following is the best explanation of the fundamental concept of uniformitarianism?**
(Rigorous) (Skill 4.2)

A. The types and varieties of life will be seen in a uniform progression over time.
B. The physical, chemical and biological laws that operated in the geologic past operate in the same way today.
C. Debris from catastrophic events (i.e. volcanoes, and meteorites) will be evenly distributed over the effected area.
D. The frequency and intensity of major geologic events will remain consistent over long periods of time.

98. **The best preserved animal remains have been discovered in _____**
(Rigorous) (Skill 4.2)

A. resin
B. shale
C. tar pits
D. glacial ice

99. **The salinity of ocean water is closest to _____ .**
(Easy) (Skill 4.3)

A. 0.035 %
B. 0.5 %
C. 3.5 %
D. 15 %

100. The theory of 'sea floor spreading' explains _____.
(Average Rigor) (Skill 4.3)

A. the shapes of the continents.
B. how continents collide.
C. how continents move apart.
D. how continents sink to become part of the ocean floor.

101. The theory of 'continental drift' is supported by which of the following?
(Average Rigor) (Skill 4.3)

A. The way the shapes of South America and Europe fit together.
B. The way the shapes of Europe and Asia fit together.
C. The way the shapes of South America and Africa fit together.
D. The way the shapes of North America and Antarctica fit together.

102. Mount Kīlauea on the island of Hawaii, is a very active volcano that has continuous lava flow into the nearby ocean. What is the name of the type of shoreline created at the point where the lava flow meets the water?
(Rigorous) (Skill 4.3)

A. Stacking
B. Submerged
C. Developing
D. Emergent

103. Surface ocean currents are caused by which of the following?
(Rigorous) (Skill 4.3)

A. temperature.
B. changes in density of water.
C. wind.
D. tidal forces.

104. Which of the following instruments measures wind speed?
(Easy) (Skill 4.4)

A. A barometer.
B. An anemometer.
C. A thermometer.
D. A weather vane.

105. A closed contour line that has tiny comb-like lines along the inner edge indicates a _____
(Average Rigor) (Skill 4.4)

A. depression
B. mountain
C. valley
D. river

106. In which layer of the atmosphere would you expect most weather conditions to occur?
(Average Rigor) (Skill 4.4)

A. Troposphere
B. Thermosphere
C. Mesosphere
D. Stratosphere

107. The transfer of heat from the earth's surface to the atmosphere is called
(Rigorous) (Skill 4.4)

A. convection.
B. radiation.
C. conduction.
D. advection.

108. Which of the following causes the aurora borealis?
(Rigorous) (Skill 4.4)

A. Particles from the sun
B. Gases escaping from earth
C. Particles from the moon
D. Electromagnetic discharges from the North pole.

109. Which of the following units is not a measure of distance?
(Easy) (Skill 4.5)

A. AU (astronomical unit).
B. Light year.
C. Parsec.
D. Lunar year.

110. The phases of the moon are the result of its _____ in relation to the sun.
(Average Rigor) (Skill 4.5)

A. revolution
B. rotation
C. position
D. inclination

111. Which of the following is a correct explanation for an astronaut's 'weightlessness' while in orbit?
(Average Rigor) (Skill 4.5)

A. Astronauts continue to feel the pull of gravity in space, but they are so far from the planets that the force is small.
B. Astronauts continue to feel the pull of gravity in space, but spacecrafts have such powerful engines that those forces dominate, reducing effective weight.
C. Astronauts do not feel the pull of gravity in space, because space is a vacuum.
D. The cumulative gravitational forces that the astronaut experiences from all sources in the solar system equal a net gravitational force of zero.

112. The planet with retrograde rotation is
(Rigorous) (Skill 4.5)

A. Pluto
B. Uranus
C. Venus
D. Saturn

113. **What is the main difference between the 'condensation hypothesis' and the 'tidal hypothesis' for the origin of the solar system?**
(Rigorous) (Skill 4.5)

A. The tidal hypothesis can be tested, but the condensation hypothesis cannot.

B. The tidal hypothesis proposes a near collision of two stars pulling on each other, but the condensation hypothesis proposes condensation of rotating clouds of dust and gas.

C. The tidal hypothesis explains how tides began on planets such as Earth, but the condensation hypothesis explains how water vapor became liquid on Earth.

D. The tidal hypothesis is based on Aristotelian physics, but the condensation hypothesis is based on Newtonian mechanics.

114. **Which of the following is a true statement about radiation exposure and air travel?**
(Average Rigor) (Skill 5.1)

A. Air travel exposes humans to radiation, but the level is not significant for most people.

B. Air travel exposes humans to so much radiation that it is recommended as a method of cancer treatment.

C. Air travel does not expose humans to radiation.

D. Air travel may or may not expose humans to radiation, but it has not yet been determined.

115. **Genetic engineering has benefited agriculture in many ways. Which of the following is not one of these benefits?**
(Rigorous) (Skill 5.1)

A. Developing a bovine growth hormone to increase milk production.

B. Strains of crops have been developed to resist herbicides.

C. The development of micro-orgasms that breakdown toxic substances into harmless compounds.

D. Genetically vaccinating plants against viral attack.

116. **Which of the following is not a common type of acid found in 'acid rain' or acidified surface water?**
(Average Rigor) (Skill 5.2)

A. Nitric acid.
B. Sulfuric acid.
C. Carbonic acid.
D. Hydrofluoric acid.

117. **Which of the following activities is least likely to lead to a disasterous event?**
(Rigorous) (Skill 5.2)

A. The use of vehicles that emit greenhouse gases.
B. Housing development on potentially unstable ground.
C. The use of nuclear power plants.
D. Strip mining.

118. **Contamination may enter groundwater by which of the following means?**
(Easy) (Skill 5.3)

A. air pollution
B. leaking septic tanks
C. photochemical processes
D. sewage treatment plants.

119. **Which of the following is the most accurate definition of a non-renewable resource?**
(Average Rigor) (Skill 5.3)

A. A non-renewable resource is never replaced once used.
B. A non-renewable resource is replaced on a timescale that is very long relative to human life-spans.
C. A non-renewable resource is a resource that can only be manufactured by humans.
D. A non-renewable resource is a species that has already become extinct.

120. **All of the following are potential sources of alternative energy that are currently being used except**
(Rigorous) (Skill 5.3)

A. Biomass energy
B. Geothermal energy
C. Wind energy
D. Nuclear fusion power plants

121. **Which of the following is not considered ethical behavior for a scientist?**
(Easy) (Skill 5.4)

A. Citing the sources before data is published.
B. Publishing data before other scientists have had a chance to replicate results.
C. Collaborating with other scientists from different laboratories.
D. Publishing work with an incomplete list of citations.

122. **Which of the following is the least ethical choice for a school laboratory activity?**
(Average Rigor) (Skill 5.4)

A. A genetics experiment tracking the fur color of mice.
B. Dissection of a preserved fetal pig.
C. Measurement of goldfish respiration rates at different temperatures.
D. Pithing a frog to observe the circulatory system.

123. **Which of the following lines of scientific research has not been strictly regulated by the federal government?**
(Rigorous) (Skill 5.4)

A. Stem-cell research
B. Nuclear physics
C. Animal testing
D. Microcomputing development

124. **Extensive use of antibacterial soap has been found to increase the virulence of certain viral strains in hospitals. Which of the following might be an explanation for this phenomenon?**
(Average Rigor) (Skill 5.5)

A. Antibacterial soaps do not kill viruses.
B. Antibacterial soaps do not incorporate the same antibiotics used as medicine.
C. Antibacterial soaps kill a lot of bacteria, and only the hardiest ones survive to reproduce.
D. Antibacterial soaps can be very drying to the skin.

125. **Which is not an example of how the general public is learning more about health issues?**
(Rigorous) (Skill 5.5)

A. Childhood Vaccinations
B. Nutrition Education
C. Radical diet plans
D. Education about signs and symptoms of heart attacks.

Answer Key

1.	B	33.	C	65.	B	97.	B
2.	D	34.	A	66.	B	98.	C
3.	A	35.	A	67.	A	99.	C
4.	B	36.	C	68.	B	100.	C
5.	B	37.	C	69.	A	101.	C
6.	B	38.	C	70.	D	102.	D
7.	A	39.	B	71.	A	103.	C
8.	B	40.	D	72.	B	104.	B
9.	B	41.	A	73.	C	105.	A
10.	B	42.	A	74.	B	106.	A
11.	B	43.	C	75.	C	107.	C
12.	A	44.	D	76.	A	108.	A
13.	B	45.	B	77.	B	109.	D
14.	D	46.	A	78.	A	110.	C
15.	B	47.	C	79.	B	111.	A
16.	A	48.	C	80.	D	112.	C
17.	B	49.	B	81.	C	113.	B
18.	C	50.	C	82.	B	114.	A
19.	D	51.	C	83.	C	115.	C
20.	C	52.	B	84.	B	116.	D
21.	A	53.	B	85.	A	117.	C
22.	C	54.	B	86.	B	118.	B
23.	C	55.	A	87.	C	119.	B
24.	B	56.	B	88.	D	120.	D
25.	C	57.	A	89.	A	121.	D
26.	C	58.	D	90.	C	122.	D
27.	B	59.	C	91.	D	123.	D
28.	C	60.	C	92.	B	124.	C
29.	C	61.	C	93.	D	125.	C
30.	B	62.	C	94.	B		
31.	B	63.	A	95.	D		
32.	A	64.	C	96.	D		

Rigor Table

	Easy %20	Average Rigor %40	Rigorous %40
Question #	1, 5, 9, 14, 19, 24, 28, 37, 42, 46, 51, 55, 60, 72, 75, 79, 82, 85, 89, 94, 99, 104, 109, 118, 121	2, 3, 6, 10, 11, 15, 16, 20, 21, 25, 26, 29, 30, 33, 34, 38, 39, 43, 47, 48, 51, 56, 57, 61, 62, 64, 65, 68, 69, 73, 76, 77, 80, 83, 86, 90, 91, 95, 96, 100, 101, 105, 106, 110, 111, 114, 116, 119, 122, 124	4, 7, 8, 12, 13, 17, 18, 22, 23, 27, 31, 32, 35, 36, 40, 41, 44, 45, 49, 50 ,53, 54, 58, 59, 63, 66, 67, 70, 71, 74, 78, 81, 84, 87, 88, 92, 93, 97, 98, 102, 103, 107, 108, 112, 113, 115, 117, 120, 123, 125

Rationales with Sample Questions

1.   **Which is the correct order of the scientific method?**
       1. collecting data
       2. planning a controlled experiment
       3. drawing a conclusion
       4. hypothesizing a result
       5. re-visiting a hypothesis to answer a question
       (Easy) (Skill 1.1)

a.   1,2,3,4,5
b.   4,2,1,3,5
c.   4,5,1,3,2
d.   1,3,4,5,2

Answer: b. 4,2,1,3,5

The correct methodology for the scientific method is first to make a meaningful hypothesis (educated guess), then plan and execute a controlled experiment to test that hypothesis. Using the data collected in that experiment, the scientist then draws conclusions and attempts to answer the original question related to the hypothesis. This is consistent only with answer (B).

2.   **For her first project of the year, a student is designing a science experiment to test the effects of light and water on plant growth. You should recommend that she _____.**
       (Average Rigor) (Skill 1.1)

a.   manipulate the temperature also.
b.   manipulate the water pH also.
c.   determine the relationship between light and water unrelated to plant growth.
d.   omit either water or light as a variable.

Answer: d. Omit either water or light as a variable.

As a science teacher for middle-school-aged students, it is important to reinforce the idea of 'constant' vs. 'variable' in science experiments. At this level, it is wisest to have only one variable examined in each science experiment. (Later, students can hold different variables constant while investigating others.) Therefore it is counterproductive to add other variables (answers (A) or (B)). It is also irrelevant to determine the light-water interactions aside from plant growth (C). So the only possible answer is (D).

3.   When designing a scientific experiment, a student considers all the factors that may influence the results. The process goal is to

_____.

(Average Rigor) (Skill 1.1)

a.   recognize and manipulate independent variables.
b.   recognize and record independent variables.
c.   recognize and manipulate dependent variables.
d.   recognize and record dependent variables.

Answer: a. Recognize and manipulate independent variables.

When a student designs a scientific experiment, s/he must decide what to measure, and what independent variables will play a role in the experiment. S/he must determine how to manipulate these independent variables to refine his/her procedure and to prepare for meaningful observations. Although s/he will eventually record dependent variables (D), this does not take place during the experimental design phase. Although the student will likely recognize and record the independent variables (B), this is not the process goal, but a helpful step in manipulating the variables. It is unlikely that the student will manipulate dependent variables directly in his/her experiment (C), or the data would be suspect. Thus, the answer is (A).

4. **In an experiment measuring the inhibition of different antibiotics on bacteria grown in Petri dishes, what are the independent and dependent variables respectively?**
(Rigorous) (Skill 1.1)

a.   Number of bacterial colonies and the antibiotic type.
b.   Antibiotic type and the distance between antibiotic and the closest colony.
c.   Antibiotic type and the number of bacterial colonies.
d.   Presence of bacterial colonies and the antibiotic type.

Answer: b. Antibiotic type and the distance between antibiotic and the closest colony.

To answer this question, recall that the independent variable in an experiment is the entity that is changed by the scientist, in order to observe the effects the dependent variable. In this experiment, the antibiotic used is purposely changed so it is the independent variable. Answers A and D list antibiotic type as the dependent variable and thus cannot be the answer, leaving answers B and C as the only two viable choices. The best answer is B, because it measures the concentration of the antibiotic at which the bacteria are able to grow (as you move from the source of the antibiotic the concentration decreases). Answer C is not the correct choice because it could be interpreted that a plate that shows a large number of colonies a greater distance from the antibiotic is a less effective antibiotic than a plate a smaller number of colonies in close proximity to the antibiotic disc, which is reverse of the actually result.

5.   Under a microscope with 440 X magnification , an object with
        diameter 0.1 millimeter appears to have a diameter of _____ .
        (Easy) (Skill 1.2)

a.   4.4 millimeters.
b.   44 millimeters.
c.   440 millimeters.
d.   4400 millimeters.

Answer: b. 44 millimeters.

To answer this question, recall that to calculate the magnified dimensions of an
object, you multiply the actual dimensions by the magnification power of the
instrument. Therefore, the 0.1 millimeter diameter is multiplied by 440. This
equals 44, so the image appears to be 44 millimeters in diameter. You could also
reason that since a 440 power microscope is considered a "high power"
microscope, you would expect a 0.1 millimeter object to appear a few
centimeters long. Therefore, the correct answer is (B).

6.   When measuring the volume of water in a graduated cylinder, where
        does one read the measurement?
        (Average Rigor) (Skill 1.2)

a.   At the highest point of the liquid.
b.   At the bottom of the meniscus curve.
c.   At the closest mark to the top of the liquid
d.   At the top of the plastic safety ring.

Answer: b. At the bottom of the meniscus curve.

To measure water in glass, you must look at the top surface at eye-level, and
ascertain the location of the bottom of the meniscus (the curved surface at the
top of the water). The meniscus forms because water molecules adhere to the
sides of the glass, which is a slightly stronger force than their cohesion to each
other. This leads to a U-shaped top of the liquid column, the bottom of which
gives the most accurate volume measurement. (Other liquids have different
forces, e.g. mercury in glass, which has a convex meniscus.) This is
consistent only with answer (B).

7. **Which of the following is not an acceptable way for a student to acknowledge sources in a laboratory report?**
   (Rigorous) (Skill 1.2)

   a. The student tells his/her teacher what sources s/he used to write the report.
   b. The student uses footnotes in the text, with sources cited, but not in correct MLA format.
   c. The student uses endnotes in the text, with sources cited, in correct MLA format.
   d. The student attaches a separate bibliography, noting each use of sources.

   Answer: a. The student tells his/her teacher what sources s/he used to write the report.

   It may seem obvious, but students are often unaware that scientists need to cite all sources used. For the young adolescent, it is not always necessary to use official MLA format (though this should be taught at some point). Students may properly cite references in many ways, but these references must be in writing, with the original assignment. Therefore, the answer is (A).

8. A scientist exposes mice to cigarette smoke, and notes that their lungs develop tumors. Mice that were not exposed to the smoke do not develop as many tumors. Which of the following conclusions may be drawn from these results?:
I. Cigarette smoke causes lung tumors.
II. Cigarette smoke exposure has a positive correlation with lung tumors in mice.
III. Some mice are predisposed to develop lung tumors.
IV. Cigarette smoke exposure has a positive correlation with lung tumors in humans.
(Rigorous) (Skill 1.2)

a. I and II only.
b. II only.
c. I , II, III and IV.
d. II and IV only.

Answer: b. II only.

Although cigarette smoke has been found to cause lung tumors (and many other problems), this particular experiment shows only that there is a positive correlation between smoke exposure and tumor development in these mice. It may be true that some mice are more likely to develop tumors than others, which is why a control group of identical mice should have been used for comparison. Mice are often used to model human reactions, but this is as much due to their low financial and emotional cost as it is due to their being a "good model" for humans, and thus this scientist cannot make the conclusion that cigarette smoke exposure has a positive correlation with lung tumors in humans based on this data alone. Therefore, the answer must be (B).

9. **Which of these is the best example of 'negligence'?**
   (Easy) (Skill 1.3)

a. A teacher fails to give oral instructions to those with reading disabilities.
b. A teacher fails to exercise ordinary care to ensure safety in the classroom.
c. A teacher does not supervise a large group of students.
d. A teacher reasonably anticipates that an event may occur, and plans accordingly.

Answer: b. A teacher fails to exercise ordinary care to ensure safety in the classroom.

'Negligence' is the failure to "exercise ordinary care" to ensure an appropriate and safe classroom environment. It is best for a teacher to meet all special requirements for disabled students, and to be good at supervising large groups. However, if a teacher can prove that s/he has done a reasonable job to ensure a safe and effective learning environment, then it is unlikely that she/he would be found negligent. Therefore, the answer is (B).

10. **Who should be notified in the case of a serious chemical spill?**
    (Average Rigor) (Skill 1.3)

a. The custodian.
b. The fire department or their municipal authority.
c. The science department chair.
d. The School Board.

Answer: b. The fire department or other municipal authority.

Although the custodian may help to clean up laboratory messes, and the science department chair should be involved in discussions of ways to avoid spills, a serious chemical spill may require action by the fire department or other trained emergency personnel. It is best to be safe by notifying them in case of a serious chemical accident. Therefore, the best answer is (B).

11. **Formaldehyde should not be used in school laboratories for the following reason:**
(Average Rigor) (Skill 1.3)

a. it smells unpleasant.
b. it is a known carcinogen.
c. it is expensive to obtain.
d. it is explosive.

Answer: b. It is a known carcinogen.

Formaldehyde is a known carcinogen, so it is too dangerous for use in schools. In general, teachers should not use carcinogens in school laboratories. Although formaldehyde also smells unpleasant, a smell alone is not a definitive marker of danger. For example, many people find the smell of vinegar to be unpleasant, but vinegar is considered a very safe classroom/laboratory chemical. Furthermore, some odorless materials are toxic. Formaldehyde is neither particularly expensive nor explosive. Thus, the answer is (B).

12. **Experiments may be done with any of the following animals except**
_____ .
(Rigorous) (Skill 1.3)

a. birds.
b. invertebrates.
c. lower order life.
d. frogs.

Answer: a. Birds.

No dissections may be performed on living mammalian vertebrates or birds. Lower order life and invertebrates may be used. Biological experiments may be done with all animals except mammalian vertebrates or birds. Therefore the answer is (A).

13. **In a science experiment, a student needs to repeatedly dispense very small measured amounts of liquid into a well-mixed solution. Which of the following is the best choice of equipment for the student?** *(Rigorous) (Skill 1.3)*

a. Pipette, Stirring Rod, Beaker.
b. Burette with Burette Stand, Stir-plate, Beaker.
c. Volumetric Flask, Dropper, Stirring Rod.
d. Beaker, Graduated Cylinder, Stir-plate.

Answer: b. Burette with Burette Stand, Stir-plate, Beaker.

The most accurate and convenient way to repeatedly dispense small measured amounts of liquid in the laboratory is with a burette, on a burette stand. To keep a solution well-mixed, a magnetic stir-plate is the most sensible choice, and the solution will usually be mixed in a beaker. Although other combinations of materials could be used for this experiment, choice (B) describes the simplest and best method. Choices A and C describe relying on a stirring a rod, which requires the student to try to mix the solution by hand while adding the liquid. Answer D describes using a graduated cylinder to add the small amounts of liquid to the solution, it is more difficult to add small quantities of a liquid with a graduated cylinder.

14. **A boulder sitting on the edge of a cliff has which type of energy?** *(Easy) (Skill 2.1)*

a. Kinetic energy
b. Latent Energy
c. No energy
d. Potential Energy

Answer: d. Potential Energy

Answer (A) would be true if the boulder fell off the cliff and started falling. Answer (C) would be a difficult condition to find since it would mean that no outside forces where operating on an object, and gravity is difficult to avoid. Answer (B) might be a good description of answer (D) which is the correct energy. The boulder has potential energy is imparted from the force of gravity.

15. **What is the definition of specific gravity?**
(Average Rigor) (Skill 2.1)

a. The mass of an object.
b. The ratio of the density of a substance to the density of water.
c. The measurement of how much mass is in a given volume.
d. The pull of the earth's gravity on an object.

Answer: b. The ratio of the density of a substance to the density of water.

Mass measures the amount of matter in an object. Density is the mass of a substance contained per unit of volume. Weight is the measure of the earth's pull of gravity on an object. The only option here is the ratio of the density of a substance to the density of water, answer (B).

16. **Which of the following statements is most accurate?**
(Average Rigor) (Skill 2.1)

a. Mass is always constant; Weight may vary by location.
b. Mass and Weight are both always constant.
c. Weight is always constant; Mass may vary by location.
d. Mass and Weight may both vary by location.

Answer: a. Mass is always constant; Weight may vary by location.

When considering chemical reactions and systems, mass is constant (mass, the amount of matter in a system, is conserved). Weight, on the other hand, is the force of gravity on an object, which is subject to change due to changes in the gravitational field and/or the location of the object. Thus, the best answer is (A).

17. **Which of the following statements describes an isotope of an element?**
(Rigorous) (Skill 2.1)

a. An isotope has a different number of electrons.
b. An isotope has a different number of neutrons.
c. The arrangement of the electrons is different.
d. An isotope has a different number of protons.

.

Answer: b. The number of Neutrons.

A change in the number of electrons (answer (A) creates an ion. The change in the arrangement of the electrons (answer (C), could change the reactivity of an atom temporary. A change of the number of Protons (answer (D), will change the atom a new element. Answer (B) is the only one that does not change the relative charge of an atom, while changing the weight of and atom, which in essence is what an isotope is.

18. **Physical properties are observable characteristics of a substance in its natural state. Which of the following are considered physical properties.**
   I   Color
   II Density
   III Specific gravity
   IV Melting Point
   (Rigorous) (Skill 2.1)

a. I only
b. I and II only
c. I, II, and III only
d. III and IV only

Answer: c. I, II, and III only

Of the possibilities only the melting point of a substance cannot be found without altering the substance itself. Color is readily observable. Density can be measured without changing a substances form or structure, and specific gravity is a ratio based on density, so once one is know the other can be calculated. Thus answer (C) is the only possible answer.

19. **The transfer of heat by electromagnetic waves is called _____ .**
    (Easy) (Skill 2.2)

a. conduction.
b. convection.
c. phase change.
d. radiation.

Answer: d. Radiation.

Heat transfer via electromagnetic waves (which can occur even in a vacuum) is called radiation. (Heat can also be transferred by direct contact (conduction), by fluid current (convection), and by matter changing phase, but these are not relevant here.) The answer to this question is therefore (D).

20. **The Law of Conservation of Energy states that _____ .**
    (Average Rigor) (Skill 2.2)

a. there must be the same number of products and reactants in any chemical equation.
b. mass and energy can be interchanged.
c. energy is neither created nor destroyed, but may change form.
d. one form energy must remain intact (or conserved) in all reactions

Answer: c. Energy is neither created nor destroyed, but may change form.

Answer (C) is a summary of the Law of Conservation of Energy (for non-nuclear reactions). In other words, energy can be transformed into various forms such as kinetic, potential, electric, or heat energy, but the total amount of energy remains constant. Answer (A) is untrue, as demonstrated by many synthesis and decomposition reactions. Answers (B) and (D) may be sensible, but they are not relevant in this case. Therefore, the answer is (C).

21. A long silver bar has a temperature of 50 degrees Celsius at one end and 0 degrees Celsius at the other end. The bar will reach thermal equilibrium (barring outside influence) by the process of:-

_____.
(Average Rigor) (Skill 2.2)

a.   heat conduction.
b.   heat convection.
c.   heat radiation.
d.   heat phase change.

Answer: a. conduction.

Heat conduction is the process of heat transfer via solid contact. The molecules in a warmer region vibrate more rapidly, jostling neighboring molecules and accelerating them. This is the dominant heat transfer process in a solid with no outside influences. Recall, also, that convection is heat transfer by way of fluid currents; radiation is heat transfer via electromagnetic waves; phase change can account for heat transfer in the form of shifts in matter phase. The answer to this question must therefore be (A).

22. **When you step out of the shower, the floor feels colder on your feet than the bathmat. Which of the following is the correct explanation for this phenomenon?**
(Rigorous) (Skill 2.2)

a. The floor is colder than the bathmat.
b. The bathmat is smaller than the floor and quickly reaches equilibrium with your body temperature.
c. Heat is conducted more easily to the floor.
d. Water is absorbed from your feet onto the bathmat so it doesn't evaporate as quickly as it does off the floor thus not cooling the bathmat as quickly.

Answer: c. Heat is conducted more easily into the floor.

When you step out of the shower and onto a surface, the surface is most likely at room temperature, regardless of its composition (eliminating answer (A)). The bathmat is likely a good insulator and is unlikely to reach equilibrium with your body temperature after a short exposure so answer (B) is incorrect. Although evaporation does have a cooling effect, it the short time it takes you to step from the bathmat to the floor, it is unlikely to have a significant effect on the floor temperature (eliminating answer (D)). Your feet feel cold when heat is transferred from them to the surface, which happens more easily on a hard floor than a soft bathmat. This is because of differences in specific heat (the energy required to change temperature, which varies by material). Therefore, the answer must be (C), i.e. heat is conducted more easily into the floor from your feet.

23. **What is the best explanation of the term "latent heat"?**
    (Rigorous) (Skill 2.2)

a.   The amount of heat it takes to change a solid to a liquid.
b.   The amount of heat being radiated by an object.
c.   The amount heat required for a substance to undergo a phase change.
d.   The amount of heat required to change a liquid to a gas.

Answer: c. The amount heat needed to change a substance to undergo a phase change.

Answer (A) is a description of the term ' heat of fusion' and answer (D) is a description of the term 'heat of vaporization', both of which a specific examples of latent heat. Answer (C) includes both of these examples by using the term 'phase change' which includes the changes from solid to liquid, and liquid to gas. Answer (B) talks about at objects giving off heat without an accompanying change in state.

24. **Which parts of an atom are located inside the nucleus?**
    (Easy) (Skill 2.3)

a.   electrons and neutrons.
b.   protons and neutrons.
c.   protons only.
d.   neutrons only.

Answer: b. Protons and Neutrons.

Protons and neutrons are located in the nucleus, while electrons move around outside the nucleus. This is consistent only with answer (B).

25. **What is the main obstacle to using nuclear fusion to generate electricity?**
    (Average Rigor) (Skill 2.3)

a. Nuclear fusion produces much more pollution than nuclear fission.
b. There is no obstacle; most power plants us nuclear fusion today.
c. Nuclear fusion requires very high temperature and activation energy.
d. The fuel for nuclear fusion is extremely expensive.

Answer: c. Nuclear fusion requires very high temperature and activation energy.

Nuclear fission is the usual process for power generation in nuclear power plants. This is carried out by splitting nuclei to release energy. The sun's energy is generated by nuclear fusion, i.e. combination of smaller nuclei into a larger nucleus. Fusion creates much less radioactive waste, but it requires extremely high temperature and activation energy, so it is not yet feasible for electricity generation. Therefore, the answer is (C).

26. **The electrons in an atom that form chemical bonds are called**

    _____.
    (Average Rigor) (Skill 2.3)

a. outer shell electrons
b. excited electrons
c. valence electrons.
d. reactive electrons.

Answer: c. valence electrons.

Answers (A), (B), and (C) could all be used as turns to describe the electrons involved in chemical bonding, depending on the situation. However only answer (C), valence electrons, is the correct answer because it the specific name given to these electron and not a description of the electrons.

27. In a fission reactor, "heavy water" is used to _____ .
(Rigorous) (Skill 2.3)

a. terminate fission reactions.
b. slow down neutrons and moderate reactions.
c. rehydrate the chemicals.
d. initiate a chain reaction.

Answer: b. Slow down neutrons and moderate reactions.

"Heavy water" is used in a nuclear [fission] reactor to slow down neutrons, controlling and moderating the nuclear reactions. It does not terminate the reaction, and it does not initiate the reaction. Also, although the reactor takes advantage of water's other properties (e.g. high specific heat for cooling), the water does not "rehydrate" the chemicals. Therefore, the answer is (B).

28. All of the following are considered Newton's Laws except for:
(Easy) (Skill 2.4)

a. An object in motion will continue in motion unless acted upon by an outside force.
b. For every action force, there is an equal and opposite reaction force.
c. Nature abhors a vacuum.
d. Mass can be considered the ratio of force to acceleration.

Answer: c. Nature abhors a vacuum.

Newton's Laws include his law of inertia (A), (an object in motion (or at rest) will stay in motion (or at rest) until acted upon by an outside force), (D) his law that (Force)=(Mass)(Acceleration), (B) and his equal and opposite reaction force law. Therefore, the answer to this question is (C), because "Nature abhors a vacuum" is not one of these.

29. All of the following are units of energy except _____
     (Average Rigor) (Skill 2.4)

a.   joules.
b.   calories.
c.   watts.
d.   ergs.

Answer: c. Watts.

Energy units must be dimensionally equivalent to (force)x(length), which equals (mass)x(length squared)/(time squared). Joules, Calories, and Ergs are all metric measures of energy. Joules are the SI units of energy, while
Calories are used to allow water to have a Specific Heat of one unit. Ergs are used in the 'cgs' (centimeter-gram-second) system, for smaller quantities. Watts, however, are units of power, i.e. Joules per Second. Therefore, the answer is (C).

30. The force of gravity on earth causes all bodies in free fall to

     _____ .

     (Average Rigor) (Skill 2.4)

a.   fall at the same speed.
b.   accelerate at the same rate.
c.   reach the same terminal velocity.
d.   move in the same direction.

Answer: b. Accelerate at the same rate.

Gravity causes approximately the same acceleration on all falling bodies close to earth's surface. (It is only "approximately" because there are very small variations in the strength of earth's gravitational field.) More massive bodies continue to accelerate at this rate for longer, before their air resistance is great enough to cause terminal velocity, so answers (A) and (C) are eliminated. Bodies on different parts of the planet move in different directions (always toward the center of mass of earth), so answer (D) is eliminated. Thus, the answer is (B).

31. **Newton's Laws are taught in science classes because _____.**
    (Rigorous) (Skill 2.4)

a.   they are the correct analysis of inertia, gravity, and forces.
b.   they are a close approximation to correct physics, for usual Earth conditions.
c.   they accurately incorporate relativity into the study of force.
d.   Newton was a well-respected scientist in his time and his ideas have stood the test of time.

Answer: b. They are a close approximation to correct physics, for usual Earth conditions.

Although Newton's Laws are often thought of as fully correct for inertia, gravity, and forces, it is important to realize that Einstein's work (and that of others) has indicated that Newton's Laws are reliable only at speeds much lower than that of light. This is reasonable, though, for most middle- and high-school applications. At speeds close to the speed of light, relativity considerations must be used. Therefore, the only correct answer is (B).

32.   Since ancient times, people have been enthralled with bird flight. What is the key to bird flight?
        (Rigorous) (Skill 2.4)

a.   Birds' wings are a particular shape and composition, that causes the air flow over the wing to travel faster than the air flow under the wing.
b.   Birds flap their wings quickly enough to propel themselves.
c.   Birds' wings are a particular shape and composition, that causes the air flow under the wing to travel faster than the air flow over the wing.
d.   Birds flap their wings to create a downward force that opposes gravity.

Answer: a. Bird wings are a particular shape and composition, that causes the air flow over the wing to travel faster than the air flow under the wing.

Birds' wings are composed of very light bones, and feathers. The shape of the feathers along the wing creates a curved upper surface and a flatter lower surface. The curved upper surface causes the air traveling over it to move faster than the air traveling under the wing because the air has a greater distance to travel. Bernoulli developed the principle that when air traveling over two sides of an object are traveling at different speeds, a lifting force is created on the side that has the faster moving airflow. So, the structure of the birds' wings creates a lifting force. Without the proper structure no amount of flapping will bring about flight, if you doubt this just try flapping your own arms. Although flapping of wings does create some downward force this force is not sufficient to get a bird off the ground, without the lifting forces that the wing structure creates. Thus, the answer is (A).

33.   Resistance is measured in units called _____ .
        (Average Rigor) (Skill 2.5)

a.   watts.
b.   volts.
c.   ohms.
d.   current.

Answer: c. Ohms.

A watt is a unit of energy. Potential difference is measured in a unit called the volt. Current is the number of electrons per second that flow past a point in a circuit. An ohm is the unit for resistance. The correct answer is (C).

34. **The electromagnetic radiation with the longest wave length is/are**
   _____.
   (Average Rigor) (Skill 2.5)

a. radio waves.
b. red light.
c. X-rays.
d. ultraviolet light.

Answer: a. Radio waves.

As one can see on a diagram of the electromagnetic spectrum, radio waves have longer wave lengths (and smaller frequencies) than visible light, which in turn has longer wave lengths than ultraviolet or X-ray radiation. If you did not remember this sequence, you might recall that wave length is inversely proportional to frequency, and that radio waves are considered much less harmful (less energetic, i.e. lower frequency) than ultraviolet or X-ray radiation. The correct answer is therefore (A).

35. **A 10 ohm resistor and a 50 ohm resistor are connected in parallel. If the current in the 10 ohm resistor is 5 amperes, the current (in amperes) running through the 50 ohm resistor is**
   (Rigorous) (Skill 2.5)

a. 1
b. 50
c. 25
d. 60

Answer: a. 1

To answer this question, use Ohm's Law, which relates voltage to current and resistance:
$V = IR$ where V is voltage; I is current; R is resistance.
We also use the fact that in a parallel circuit, the voltage is the same across the branches. Because we are given that in one branch, the current is 5 amperes and the resistance is 10 ohms, we deduce that the voltage in this circuit is their product, 50 volts (from $V = IR$). We then use $V = IR$ again, this time to find I in the second branch. Because V is 50 volts, and R is 50 ohm, we calculate that I has to be 1 ampere. This is consistent only with answer (A).

36. **A light bulb is connected in series with a rotating coil within a magnetic field. The brightness of the light may be increased by any of the following except:**
(Rigorous) (Skill 2.5)

a. Rotating the coil more rapidly.
b. Adding more loops to the coil.
c. Using tighter loops for the coil.
d. Using a stronger magnetic field.

Answer: c. Using a tighter loops for the coil.

To answer this question, recall that the rotating coil in a magnetic field generates electric current, by Faraday's Law. Faraday's Law states that the amount of emf generated is proportional to the rate of change of magnetic flux through the loop. This increases if the coil is rotated more rapidly (A), if there are more loops (B), or if the magnetic field is stronger (D). Tighter loops would not change the amount of material in the loops, like using more loops would

37. **Sound waves are produced by _____ .**
(Easy) (Skill 2.6)

a. pitch.
b. noise.
c. vibrations.
d. sonar.

Answer: c. vibrations.

Sound waves are produced by a vibrating body. The vibrating object moves forward and compresses the air in front of it, then reverses direction so that the pressure on the air is reduced and expansion of the air molecules occurs. The vibrating air molecules move back and forth parallel to the direction of motion of the wave as they pass the energy from adjacent air molecules closer to the source to air molecules farther away from the source. Therefore, the answer is (C).

38. **The Doppler Effect is most closely with which property of waves?**
    (Average Rigor) (Skill 2.6)

a. amplitude.
b. wavelength.
c. frequency.
d. intensity.

Answer: c. Frequency.

The Doppler Effect accounts for an apparent increase in frequency when a wave source moves toward a wave receiver or apparent decrease in frequency when a wave source moves away from a wave receiver. (Note that the receiver could also be moving toward or away from the source.) As the wave fronts are released, motion toward the receiver mimics more frequent wave fronts, while motion away from the receiver mimics less frequent wave fronts. Meanwhile, the amplitude, wavelength, and intensity of the wave are not as relevant to this process (although moving closer to a wave source makes it seem more intense). The answer to this question is therefore (C).

39. **The speed of light changes in different materials. This is due to**
    _____ .
    (Average Rigor) (Skill 2.6)

a. interference.
b. refraction.
c. reflection.
d. relativity.

Answer: b. Refraction.

Refraction (B) is the bending of light because it hits a material at an angle which results in a change in its speed. (This is analogous to a cart rolling on a smooth road. If it hits a rough patch at an angle, the wheel on the rough patch slows down first, leading to a change in direction). Interference (A) occurs when light waves interfere with each other to form brighter or dimmer patterns; reflection (C) occurs when light bounces off a surface; relativity (D) is a general topic related to light speed and its implications, but not specifically indicated here. Therefore, the answer is (B).

40.   A converging lens produces a real image _____.
        (Rigorous) (Skill 2.6)

a.   never
b.   when the object is exactly at a distance of one focal length.
c.   when the object is within one focal length of the lens.
d.   when the object is further than one focal length from the lens.

Answer: d. when the object is further than one focal length from the lens.

Converging lens produce real images when the object is far enough from the lens (outside one focal length) so that the rays of light that bounce off the object can hit the lens and be focused into a real image on the other side of the lens. When the object is closer than one focal length from the lens, rays of light do not converge on the other side; they diverge. This means that only a virtual image can be formed, i.e. the theoretical place where those diverging rays would have converged if they had originated behind the object. Thus, the correct answer is (D).

41.   **As a train approaches and blasts its whistle, the whistle sounds**

        _____ .
        (Rigorous) (Skill 2.6)

a.   higher, because it has a higher apparent frequency.
b.   lower, because it has a lower apparent frequency.
c.   higher, because it has a lower apparent frequency.
d.   lower, because it has a higher apparent frequency.

Answer: a. Higher, because it has a higher apparent frequency.

By the Doppler effect, when a source of sound is moving toward an observer, the wave fronts are closer together, i.e. they have a greater apparent frequency. Higher frequency sounds are higher in pitch. This is consistent only with answer (A).

42.   The elements in the modern Periodic Table are arranged _____.
       (Easy) (Skill 2.7)

a.   in numerical order by atomic number.
b.   randomly.
c.   in alphabetical order by chemical symbol.
d.   in the order of their discovery.

Answer: a. In numerical order by atomic number.

The modern table is arranged by atomic number, i.e. the number of protons in each element. (This allows the element list to be complete and unique). The elements are not arranged either randomly or in alphabetical order. The answer to this question is therefore (A).

43.   Which group of metals is the most chemically active?
       (Average Rigor) (Skill 2.7)

a.   Alkaline Earth Metals
b.   Transition elements
c.   Alkali Metals
d.   Metalloids

Answer: c. Alkali Metals

Answer (D) is the only answer of the four that is not considered one of the metal groups and can thus be dismissed. From right to left on the periodic table the answers would be (C), (A), and then (B), this is also the order of decreasing chemical activity that we see. Thus the answer is (C). An impressive, yet somewhat dangerous demonstration of the Alkali metals activity can be seen by the small explosion that is created when a small amount of pure sodium is dropped into water.

44. **Which of the following is not a property of metalloids?**
(Rigorous) (Skill 2.7)

a.   Metalloids are solids at standard temperature and pressure.
b.   Metalloids can conduct electricity to a limited extent.
c.   Metalloids are found in groups 13 through 17.
d.   Metalloids all favor ionic bonding.

Answer: d. Metalloids all favor ionic bonding.

Metalloids are substances that have characteristics of both metals and nonmetals, including limited conduction of electricity and solid phase at standard temperature and pressure. Metalloids are found in a 'stair-step' pattern from Boron in group 13 through Astatine in group 17. Some metalloids, e.g. Silicon, favor covalent bonding. Others, e.g. Astatine, can bond ionically. Therefore, the answer is (D). Recall that metals/nonmetals/metalloids are not strictly defined by Periodic Table group, so their bonding is unlikely to be consistent with one another.

45. **Which of the following statements are true of all transition elements?**
(Rigorous) (Skill 2.7)

a.   They are all hard solids at room temperature.
b.   They tend to form salts when reacted with Halogens.
c.   They all have a silvery appearance in their pure state.
d.   All of the Above

Answer: b. They tend to form salts when reacted with Halogens.

Answer (A) is incorrect because Mercury which has a low melting point is a liquid at room temperature. Answer (C) is incorrect because Copper and Gold do not have a silvery appearance in the natural states. Since answers (A) and (C) are not correct then answer (D) cannot be correct either. This leaves only answer (B).

46. **What is the best definition of Isomerization?**
(Easy) (Skill 2.8)

a.   A chemical reaction in which a molecule changes shape, but no atoms are lost or gained.
b.   A chemical reaction in which a molecule changes shape, but one or more atoms are lost.
c.   A chemical reaction in which a molecule changes shape, but one or more atoms are gained.
d.   A chemical reaction in which a compound is broken down into its constitute parts.

Answer: a. A chemical reaction in which a molecule changes shape, but no atoms are lost or gained.

Answers (B) or (D) would be applicable to a decomposition reaction, Answer (C) is likely applicable to a combination or synthesis reaction. Answer (A) best explains isomerization in which a chemical rearranges its chemical bonds (these changing shape) without losing or adding any new atoms, although how those atoms are attached to each other will be different.

47. **Which of the following is found in the least abundance in organic molecules?**
(Average Rigor) (Skill 2.8)

a.   Phosphorus.
b.   Potassium.
c.   Argon.
d.   Oxygen.

Answer: c. Argon.

Organic molecules consist mainly of Carbon, Hydrogen, and Oxygen, also contain significant amounts of Nitrogen, Phosphorus, and often Sulfur. Other elements, such as Potassium, are present in much smaller quantities. Argon is a noble gas, so the atoms rarely bond to any other atoms, making it extremely rare for argon to be part of an organic compound. Therefore the answer is (C).

48. **What is necessary for ion diffusion to occur spontaneously?** *(Average Rigor) (Skill 2.8)*

a. Carrier proteins.
b. Energy from an outside source.
c. A concentration gradient.
d. Activation Energy

Answer: c. A concentration gradient.

Spontaneous diffusion occurs when random motion leads particles to increase entropy by equalizing concentrations. Particles tend to move into places of lower concentration. Therefore, a concentration gradient is required, so the answer is (C). No proteins (A), outside energy (B), or activation energy (which is also outside energy) (D) are required for this process.

49. **The chemical equation for water formation is: $2H_2 + O_2 \rightarrow 2H_2O$. Which of the following is an incorrect interpretation of this equation?** *(Rigorous) (Skill 2.8)*

a. Two moles of hydrogen gas and one mole of oxygen gas combine to make two moles of water.
b. Two grams of hydrogen gas and one gram of oxygen gas combine to make two grams of water.
c. Two molecules of hydrogen gas and one molecule of oxygen gas combine to make two molecules of water.
d. Four atoms of hydrogen (combined as a diatomic gas) and two atoms of oxygen (combined as a diatomic gas) combine to make two molecules of water.

Answer: b. Two grams of hydrogen gas and one gram of oxygen gas combine to make two grams of water.

In any chemical equation, the coefficients indicate the relative proportions of molecules (or atoms), or of moles of molecules. They do not refer to mass, because chemicals combine in repeatable combinations of molar ratio (i.e. number of moles), but vary in mass per mole of material. Therefore, the answer must be the only choice that does not refer to numbers of particles, i.e. answer (B), which refers to grams, a unit of mass.

50. **Carbon forms bonds with hydrogen by _____ .**
    (Rigorous) (Skill 2.8)

a.   ionic bonding.
b.   non-polar covalent bonding.
c.   polar covalent bonding.
d.   strong nuclear force.

Answer: c. Polar covalent bonding.

Each carbon atom contains four valence electrons, while each hydrogen atom contains one valence electron. A carbon atom can bond with one or more hydrogen atoms, such that two electrons are shared in each bond. This is covalent bonding, because the electrons are shared. (In ionic bonding, atoms must gain or lose electrons to form ions. The ions are then electrically attracted in oppositely-charged pairs.) Covalent bonds are always polar when between two non-identical atoms, so this bond must be polar. ("Polar" means that the electrons are shared unequally, forming a pair of partial charges, i.e. poles.) In any case, the strong nuclear force is not relevant to this problem. The answer to this question is therefore (C).

51. **A cup of hot liquid and a cup of cold liquid are both sitting in a room at a temperature of 72 degrees Fahrenheit and 25% humidity. Both cups are made of thin plastic. Which of the following is a true statement?**
(Easy) (Skill 2.9)

a. There will be condensation on the outside of both cups.
b. There will be condensation on the outside of the hot liquid cup, but not on the cold liquid cup.
c. There will be condensation on the outside of the cold liquid cup, but not on the hot liquid cup.
d. There will not be condensation on the outside of either cup.

Answer: c. There will be condensation on the outside of the cold liquid cup, but not on the hot liquid cup.

Condensation forms on the outside of a cup when the contents of the cup are colder than the surrounding air, and the material the cup is made of is not a perfect insulator. This happens because the air surrounding the cup is cooled to a lower temperature than the ambient room, so it has a lower saturation point for water vapor. Although the humidity was reasonable in the warmer air, when that air circulates near the colder region and cools, water condenses onto the cup's outside surface. This phenomenon is also visible when someone takes a hot shower, and the mirror gets foggy. The mirror surface is cooler than the ambient air, and provides a surface for water condensation. Furthermore, the same phenomenon is occurs when defrosting car windows; defrosters send heat to the windows—the warmer window does not permit as much condensation.
Therefore, the correct answer is (C).

52. **When heat is added to most solids, they expand. Why is this the case?**
(Average Rigor) (Skill 2.9)

a.   The molecules become larger..
b.   The increased molecular motion leads to greater distance between the molecules.
c.   The molecules develop greater repelling electric forces when heated.
d.   The molecules form a more rigid structure.

Answer: b. The increased molecular motion leads to greater distance between the molecules.

The atomic theory of matter states that matter is made up of tiny, rapidly moving particles. These particles move more quickly when warmer, because temperature is a measure of average kinetic energy of the particles. Warmer molecules therefore move further away from each other, with enough energy to separate from each other more often and for greater distances. The individual molecules do not get bigger, by conservation of mass, eliminating answer (A). The molecules do not develop greater repelling electric forces, eliminating answer (C). Occasionally, molecules form a more rigid structure when becoming colder and freezing (such as water)—but this gives rise to the exceptions to heat expansion, so it is not relevant here, eliminating answer (D). Therefore, the answer is (B).

53. **Which of the following is not true about phase change in matter? (Rigorous) (Skill 2.9)**

a. Solid water and liquid ice can coexist at water's freezing point.
b. At 7 degrees Celsius, water is always in liquid phase.
c. Matter changes phase when enough energy is gained or lost.
d. Different phases of matter are characterized by differences in molecular motion.

Answer: b. At 7 degrees Celsius, water is always in liquid phase.

According to the molecular theory of matter, molecular motion determines the 'phase' of the matter, and the energy in the matter determines the speed of molecular motion. Solids have vibrating molecules that are in fixed relative positions; liquids have faster molecular motion than their solid forms, and the molecules may move more freely but must still be in contact with one another; gases have even more energy and more molecular motion. (Other phases, such as plasma, are yet more energetic.) At the 'freezing point' or 'boiling point' of a substance, both relevant phases may be present. For instance, water at zero degrees Celsius may be composed of some liquid and some solid, or all liquid, or all solid. Pressure changes, in addition to temperature changes, can cause phase changes. For example, nitrogen can be liquefied under high pressure, even though its boiling temperature is very low. Therefore, the correct answer must be (B). Water may be a liquid at that temperature, but it may also be a solid, depending on ambient pressure.

54. **If the volume of a confined gas is increased, what happens to the pressure of the gas? You may assume that the gas behaves ideally, and that temperature and number of gas molecules remain constant. (Rigorous) (Skill 2.9)**

a.   The pressure increases.
b.   The pressure decreases.
c.   The pressure stays the same.
d.   There is not enough information given to answer this question.

Answer: b. The pressure decreases, and the temperature decreases.

Because we are told that the gas behaves ideally, you may assume that it follows the Ideal Gas Law, i.e. $PV = nRT$. This means that an increase in volume is associated with a decrease in pressure (i.e. higher T means lower P), because we are also given that all the components of the right side of the equation remain constant. Therefore, the answer must be (B).

55. **Catalysts assist reactions by _____ .**
(Easy) (Skill 2.10)

a.   lowering required activation energy.
b.   maintaining precise pH levels.
c.   keeping systems at equilibrium.
d.   changing the starting amounts of reactants.

Answer: a. Lowering required activation energy.

Chemical reactions can be enhanced or accelerated by catalysts, which are present both with reactants and with products. They induce the formation of activated complexes, thereby lowering the required activation energy—so that less energy is necessary for the reaction to begin. Catalysts may require a well maintained pH to operate effectively, however they do not do this themselves. A catalyst, by lowering activation energy, may change a reaction's equilibrium point however it does not maintain a system at equilibrium. The starting level of reactants is controlled separately from the addition of the catalyst, and has no direct correlation. Thus the correct answer is (A).

56.   **Which of the following will not change in a chemical reaction?**
       (Average Rigor) (Skill 2.10)

a.   Number of moles of products.
b.   Atomic number of one of the reactants.
c.   Mass (in grams) of one of the reactants.
d.   Temperature of the system.

Answer: b. Atomic number of one of the reactants.

Atomic number, i.e. the number of protons in a given element, is constant unless involved in a nuclear reaction. Meanwhile, the amounts (measured in moles (A) or in grams (C) of reactants and products change over the course of a chemical reaction, and the temperature of the system (D) may change due to internal or external processes. Therefore, the answer is (B).

57.   **Which of the following is a correct definition for 'chemical equilibrium'?**
       (Average Rigor) (Skill 2.10)

a.   Chemical equilibrium occurs when the forward and backward reaction rates are equal. The reaction may continue to proceed forward and backward.
b.   Chemical equilibrium occurs when the forward and backward reaction rates are equal, and equal to zero. The reaction does not continue.
c.   Chemical equilibrium occurs when there are equal quantities of reactants and products.
d.   Chemical equilibrium occurs when acids and bases neutralize each other fully.

Answer: a. Chemical equilibrium is when the forward and backward reaction rates are equal. The reaction may continue to proceed forward and backward.

Chemical equilibrium occurs when the quantities of reactants and products are at a 'steady state' and are no longer shifting, but the reaction may still proceed forward and backward. The rate of forward reaction must equal the rate of backward reaction. Note that there may or may not be equal amounts of chemicals, and that this is not restricted to a completed reaction or to an acid-base reaction. Therefore, the answer is (A).

58. **Which change does not affect enzymatic rates?**
    (Rigorous) (Skill 2.10)

a. Increase the temperature.
b. Add more substrate.
c. Adjust the pH.
d. Use a larger cell.

Answer: d. Use a larger cell.

Temperature, chemical concentration, and pH can all affect enzymatic rates. However, the chemical reactions take place on a small enough scale that the overall cell size is not relevant. Therefore, the answer is (D).

59. **Which reaction below is a decomposition reaction?**
    (Rigorous) (Skill 2.10)

a. $HCl + NaOH \rightarrow NaCl + H_2O$
b. $C + O_2 \rightarrow CO_2$
c. $2H_2O \rightarrow 2H_2 + O_2$
d. $CuSO_4 + Fe \rightarrow FeSO_4 + Cu$

Answer: c. $2H_2O \rightarrow 2H_2 + O_2$

To answer this question, recall that a decomposition reaction is one in which there are fewer reactants (on the left) than products (on the right). This is consistent only with answer (C). Meanwhile, note that answer (A) shows a double-replacement reaction (in which two sets of ions switch bonds), answer (B) shows a synthesis reaction (in which there are fewer products than reactants), and answer (D) shows a single-replacement reaction (in which one substance replaces another in its bond, but the other does not get a new bond).

60.   Vinegar is an example of a _____ .
        (Easy) (Skill 2.11)

a.   strong acid.
b.   strong base.
c.   weak acid.
d.   weak base.

Answer: c. Weak acid.

The main ingredient in vinegar is acetic acid, a weak acid. Vinegar is a useful acid in science classes, because it makes a frothy reaction with bases such as baking soda (e.g. in the quintessential volcano model). Vinegar is not a strong acid, such as hydrochloric acid, because it does not dissociate as fully or cause as much corrosion. And it is not a base. Therefore, the answer is (C).

61.   **Which of the following statements are true of vapor pressure at equilibrium?**
        (Average Rigor) (Skill 2.11)

a.   Solids have no vapor pressure.
b.   Dissolving a solute in a liquid increases its vapor pressure.
c.   The vapor pressure of a pure substance is characteristic of that substance and its temperature.
d.   The vapor pressure of a gas is unique to each gas and is independent of temperature.

Answer: c. The vapor pressure of a pure substance is characteristic of that substance and its temperature.

Only temperature and the identity of the substance determine vapor pressure. Solids have a vapor pressure, and solutes decrease vapor pressure.

62. **Which one of the following compounds would form the strongest electrolytes when dissolved in water?**
(Average Rigor) (Skill 2.11)

a. Glucose
b. Lemon juice
c. NaBr
d. Heptane

Answer: c. NaBr

NaBr like NaCl will completely ionize in the water making it a strong electrolyte. This makes (C) the correct answer.. Lemon juice (B), an acid, will partially ionize making it a weak electrolyte. Neither Glucose (A) or Heptane (D) will ionize at all, so neither are electrolytes. Furthermore the heptane is hydrophobic and will not dissolve in the water at all.

63. **Which of the following occur when NaCl dissolves in water?**
(Rigorous) (Skill 2.11)

a. Heat is required to break bonds in the NaCl crystal lattice.
b. Heat is released when hydrogen bonds in water are broken.
c. Heat is required to form bonds of hydration.
d. The oxygen end of the water molecule is attracted to the Cl^- ion.

Answer: a. Heat is required to break bonds in the NaCl crystal lattice.

The lattice does break apart, H-bonds in water are broken, and bonds of hydration are formed, but the first and second process require heat while the third process releases heat. The oxygen end of the water molecule has a partial negative charge and is attracted to the Na^+ ion.

64. The first stage of mitosis is called _____ .
(Average Rigor) (Skill 3.1)

a. telophase.
b. anaphase.
c. prophase.
d. metaphase

Answer: c. Prophase.

In mitosis, the division of somatic cells, prophase is the stage where the cell enters mitosis. The four stages of mitosis, in order, are: prophase, metaphase, anaphase, and telophase. During prophase, the cell begins the nonstop process of division. Its chromatin condenses, its nucleolus disappears, the nuclear membrane breaks apart, mitotic spindles form, its cytoskeleton breaks down, and centrioles push the spindles apart. Note that interphase, the stage where chromatin is loose, chromosomes are replicated, and cell metabolism is occurring, is technically not a stage of mitosis; it is a precursor to cell division.

65. Which process(es) result(s) in a haploid chromosome number?
(Average Rigor) (Skill 3.1)

a. Mitosis.
b. Meiosis.
c. Both mitosis and meiosis.
d. Neither mitosis nor meiosis.

Answer: b. Meiosis.

Meiosis is the division of sex cells. The resulting chromosome number is half the number of parent cells, i.e. a 'haploid chromosome number'. Mitosis, however, is the division of somatic cells, in which the chromosome number is the same as the parent cell chromosome number. Therefore, the answer is (B).

66.   A by-product of anaerobic respiration in animals is _____.
       (Rigorous) (Skill 3.1)

a.   carbon dioxide.
b.   lactic acid.
c.   oxygen.
d.   sodium chloride

Answer: b. Lactic acid.

In animals, anaerobic respiration (i.e. respiration without the presence of oxygen) generates lactic acid as a byproduct. (Note that some anaerobic bacteria generate carbon dioxide from respiration of methane, and animals generate carbon dioxide in aerobic respiration). Oxygen is not normally a by-product of respiration, though it is a product of photosynthesis, and sodium chloride is not relevant in this question. Therefore, the answer must be (B). Lactic acid causes muscle soreness after anaerobic weight-lifting.

67.   Which cellular organelle contains the food and other materials needed by the cell?
       (Rigorous) (Skill 3.1)

a.   Vacuoles.
b.   Golgi Apparatus.
c.   Ribosomes.
d.   Lysosomes.

Answer: a. Vacuoles.

In a cell, the sub-parts are called organelles. Of these, the vacuoles hold stored food (and water and pigments). The Golgi Apparatus sorts molecules from other parts of the cell; the ribosomes are sites of protein synthesis; the lysosomes contain digestive enzymes. This is consistent only with answer (A).

68. **Which of the following is not a nucleotide?**
    (Average Rigor) (Skill 3.2)

a. adenine.
b. alanine.
c. cytosine.
d. guanine.

Answer: b. Alanine.

Alanine is an amino acid. Adenine, cytosine, guanine, thymine, and uracil are nucleotides. The correct answer is (B).

69. **A white flower is crossed with a red flower. Which of the following is a sign of incomplete dominance?**
    (Average Rigor) (Skill 3.2)

a. Pink flowers.
b. Red flowers.
c. White flowers.
d. No flowers.

Answer: a. Pink flowers.

Incomplete dominance means that neither the red nor the white gene is strong enough to suppress the other. Therefore both are expressed, leading in this case to the formation of pink flowers. Therefore, the answer is (A).

70. A child has type O+ blood. Her father has type A+ blood, and her mother has type B- blood. What are the genotypes of the father and mother, respectively?
(Rigorous) (Skill 3.2)

a. AO+ - and BO - -.
b. AO+ + and BO - -.
c. AO+ + and BO + -.
d. Cannot determine both parents genotype from the information provided

Answer: d. Cannot determine both parents genotype from the information provided

Because O blood is recessive, the child must have inherited two O's—one from each of her parents. Since her father has type A blood, his genotype must be AO; likewise her mother's blood must be BO. Because the lack of the Rh factor (-) is recessive the child must have inherited at least one Rh+ from a parent. The father contributed the Rh+ to the child however it cannot be determined if the father is heterozygous for the Rh factor, answer (A), or homozygous for the Rh factor, answer (B). Answer (C) is able to be dismissed because the mother cannot be heterozygous for the Rh factor and express the recessive trait. Since both answers (A) and (B) are possible then the answer must be (D). With additional family information the father's genotype maybe be able to be determined.

71. Amino acids are carried to the ribosome in protein synthesis by _____ .
(Rigorous) (Skill 3.2)

a. transfer RNA (tRNA).
b. transport enzymes.
c. ribosomal RNA (rRNA).
d. cytoskeletal transport proteins.

Answer: a. Transfer RNA (tRNA).

Transfer RNA (tRNA) carries and position amino acids to/on the ribosomes. Messenger (mRNA) copies DNA code and brings it to the ribosomes; rRNA is in the ribosome itself. Although there are enzymes and proteins, both attached and not attached to the cell's cytoskeleton, neither transport individual amino acids to the ribosome. Thus, the answer is (A).

72.   A duck's webbed feet are an example of _____ .
        (Easy) (Skill 3.3)

a.   mimicry.
b.   structural adaptation.
c.   protective resemblance.
d.   protective coloration.

Answer: b. Structural adaptation.

Ducks (and other aquatic birds) have webbed feet, which makes them more efficient swimmers. This is most likely due to evolutionary patterns where webbed-footed-birds were more successful at feeding and reproducing, and eventually became the majority of aquatic birds. Because this structure of the duck adapted to its environment over generations, this is termed 'structural adaptation'. Mimicry, protective resemblance, and protective coloration refer to other evolutionary mechanisms for survival. The answer to this question is therefore (B).

73.   **Which of the following is not one of the principles of Darwin's Theory of Natural Selection?**
        (Average Rigor) (Skill 3.3)

a.   More individuals are produced than will survive.
b.   The individuals in a certain species vary from generation to generation.
c.   Only the fittest members of a species survive.
d.   Some genes allow for better survival of an animal.

Answer: c. Only the fittest members of a species survive.

Answers (A), (B) and (D) were all specifically noted by Darwin in his theory. Answer (C) is often misquoted to represent this particular theory, but was not mentioned by Darwin himself.

74. **Which of the following is the best example of an explanation of the theory of evolution?**
(Rigorous) (Skill 3.3)

a. Giraffes need to reach higher for leaves to eat, so their necks stretch. The giraffe babies are then born with longer necks. Eventually, there are more long-necked giraffes in the population.

b. Giraffes with longer necks are better able to reach more leaves, so they eat more and have more babies than other giraffes. Eventually, there are more long-necked giraffes in the population.

c. Giraffes want to reach higher for leaves to eat, so they release enzymes into their bloodstream, which in turn causes fetal development of longer-necked giraffes. Eventually, there are more long-necked giraffes in the population.

d. Giraffes with long necks are more attractive to other giraffes, so they get the best mating partners and have more babies. Eventually, there are more long-necked giraffes in the population.

Answer: b. Giraffes with longer necks are better able to reach more leaves, so they eat more and have more babies than other giraffes. Eventually, there are more long-necked giraffes in the population.

Although evolution is often misunderstood, it occurs via natural selection. Organisms with a life/reproductive advantage will produce more offspring. Over many generations, this changes the proportions within the population. Nonetheless, it is impossible for a stretched neck (A) or a fervent desire (C) to result in a biologically altered baby. Although there are traits that are naturally selected because of mate attractiveness and fitness (D), this is not the primary situation here, so answer (B) is the best choice.

75.   Members of the same animal species _____ .
        (Easy) (Skill 3.4)

a.   look identical.
b.   never adapt differently.
c.   are able to reproduce with each other.
d.   are found in the same geographic location.

Answer: c. Are able to reproduce with each other.

Although members of the same animal species may look alike (A), adapt alike (B), or be found near each other (D), the only requirement is that they be able to reproduce with one another. This ability to reproduce within the group is considered the hallmark of a species. Therefore, the answer is (C).

76.   **Which of the following is not a necessary characteristic of living things?**
        (Average Rigor) (Skill 3.4)

a.   Movement.
b.   Reduction of local entropy.
c.   Composed of cells.
d.   Reproduction.

Answer: a. Movement.

There are many definitions of "life," but in all cases, a living organism reduces local entropy, changes chemical energy into other forms, and reproduces. Not all living things move, however, so the correct answer is (A).

77. **Animals with a notochord or backbone are in the phylum**
     (Average Rigor) (Skill 3.4)

a.   Arthropoda.
b.   Chordata.
c.   Mollusca.
d.   Ammalia.

Answer: b. Chordata.

The phylum arthropoda includes spiders and insects and phylum mollusca contain snails and squid. Mammalia is a class in the phylum chordata. The answer is (B).

78. **Laboratory researchers have classified fungi as distinct from plants because the cell walls of fungi _____ .**
     (Rigorous) (Skill 3.4)

a.   contain chitin.
b.   contain yeast.
c.   are more solid.
d.   are less solid.

Answer: a. Contain chitin.

Kingdom Fungi consists of organisms that are eukaryotic, multicellular, absorptive consumers. They have a chitin cell wall, which is the only universally present feature in fungi that is never present in plants. Thus, the answer is (A).

79. **Which part of a plant is responsible for transporting water?**
    (Easy) (Skill 3.5)

a.   Phloem
b.   Xylem
c.   Stomata
d.   Cortex

Answer: b. Xylem

The Phloem transport a plant's food. Stomata are openings on the underside of a leaf that allow for the passage of carbon dioxide, oxygen and water. The Cortex is where a plant stores food. The only answer is (B) the Xylem, which transports water up the plant.

80. **Which of the following organisms uses spores to reproduce?**
    (Average Rigor) (Skill 3.5)

a.   Fish.
b.   Flowering plants.
c.   Conifers.
d.   Ferns.

Answer: d. Ferns.

Ferns reproduce with spores and flagellated sperm. Flowering plants reproduce via seeds, and conifers reproduce via seeds protected in cones (e.g. pine cone). Fish, of course, reproduce sexually. Therefore, the answer is (D).

81. **Which of the following is not characteristic of Gymnosperms?**
(Rigorous) (Skill 3.5)

a. They are less dependent on water for reproduction than other plant groups.
b. Gymnosperms have cones which protect their seeds.
c. Gymnosperms reproduce asexually.
d. Gymnosperm seeds and pollen are easily carried by the wind.

Answer: c. Gymnosperms reproduce asexually.

Gymnosperms (which means naked seeds) were the first plants to evolve with seeds. They are less dependent on water to assist in reproduction, and their seeds are transported by wind. Pollen from the male is also carried by the wind. Thus, Gymnosperms cannot be asexual, which makes (C) the correct answer.

82. **Which is the correct sequence of insect development?**
(Easy) (Skill 3.6)

a. Egg, pupa, larva, adult.
b. Egg, larva, pupa, adult.
c. Egg, adult, larva, pupa.
d. Pupa, egg, larva, adult.

Answer: b. Egg, larva, pupa, adult.

An insect begins as an egg, hatches into a larva (e.g. caterpillar), forms a pupa (e.g. cocoon), and emerges as an adult (e.g. moth). Therefore, the answer is (B).

83. **Echinodermata are best known for what characteristic?**
     (Average Rigor) (Skill 3.6)

a. Their slimy skin
b. Their Dry Habitat
c. Their tube feet
d. Their tentacles.

Answer: c. Their tube feet.

Echinodermata include sea urchins and starfish. They live in marine habitats, have spiny skin, and do not have tentacles. Thus, the best known characteristic here would have to be their tube feet, which they use for locomotion and feeding.

84. **Mollusca have an open circulatory system. Their sinuses serve which purpose?**
     (Rigorous) (Skill 3.6)

a. Breathing
b. Bathing
c. Filtering food
d. Circulating blood

Answer: b. Bathing

Creatures in the Mollusca genus include clams, octupi, and soft bodied animals which have a muscular foot for movement. Most of these creatures breathe through gills. With the open circulatory system, the sinuses are for bathing the body regions of the creature.

85.   **What are the most significant and prevalent elements in the biosphere?**
       (Easy) (Skill 3.7)

a.   Carbon, Hydrogen, Oxygen, Nitrogen, Phosphorus.
b.   Carbon, Hydrogen, Sodium, Iron, Calcium.
c.   Carbon, Oxygen, Sulfur, Manganese, Iron.
d.   Carbon, Hydrogen, Oxygen, Nickel, Sodium, Nitrogen.

Answer: a. Carbon, Hydrogen, Oxygen, Nitrogen, Phosphorus.

Organic matter (and life as we know it) is based on Carbon atoms, bonded to Hydrogen and Oxygen. Nitrogen and Phosphorus are the next most significant elements, followed by Sulfur and then trace nutrients such as Iron, Sodium, Calcium, and others. Therefore, the answer is (A). If you know that the formula for any carbohydrate contains Carbon, Hydrogen, and Oxygen, that will help you narrow the choices to (A) and (D) in any case.

86.   **A wrasse (fish) cleans the teeth of other fish by eating away plaque. This is an example of _____ between the fish.**
       (Average Rigor) (Skill 3.7)

a.   parasitism.
b.   symbiosis (mutualism).
c.   competition.
d.   predation.

Answer: b. Symbiosis (mutualism).

When both species benefit from their interaction in their habitat, this is called 'symbiosis', or 'mutualism'. In this example, the wrasse benefits from having a source of food, and the other fish benefit by having healthier teeth. Note that 'parasitism' is when one species benefits at the expense of the other, 'competition' is when two species compete with one another for the same habitat or food, and 'predation' is when one species feeds on another. Therefore, the answer is (B).

87. **What is the most accurate description of the Water Cycle?**
   (Rigorous) (Skill 3.7)

a.   Rain comes from clouds, filling the ocean. The water then evaporates and becomes clouds again.
b.   Water circulates from rivers into groundwater and back, while water vapor circulates in the atmosphere.
c.   Water is conserved except for chemical or nuclear reactions, and any drop of water could circulate through clouds, rain, groundwater, and surface-water.
d.   Water flows toward the oceans, where it evaporates and forms clouds, which causes rain, which in turn flow back to the oceans after it falls.

Answer: c. Water is conserved except for during chemical or nuclear reactions; a drop of water could circulate through clouds, rain, groundwater, and surface-water.

All natural chemical cycles, including the Water Cycle, depend on the principle of Conservation of Mass. (For water, unlike for elements such as Nitrogen, chemical reactions may cause sources or sinks of water molecules.)
Any drop of water may circulate through the hydrologic system, ending up in a cloud, as rain, or as surface- or groundwater. Although answers (A), (B) and (D) describe parts of the water cycle, the most comprehensive and correct answer is (C).

88. **What is the source of drinking water for most of the United States?** *(Rigorous) (Skill 3.7)*

a. Desalinated ocean water.
b. Surface water (lakes, streams, mountain runoff).
c. Rainfall into municipal reservoirs.
d. Groundwater.

Answer: d. Groundwater.

Groundwater currently provides drinking water for 53% of the population of the United States. (Although groundwater is often less polluted than surface water, it can be contaminated and it is very hard to clean once it is polluted. If too much groundwater is used from one area, then the ground may sink or shift, or local salt water may intrude from ocean boundaries). The other answer choices can be used for drinking water, but they are not the most widely used. Therefore, the answer is (D).

89. _____ **are areas of weakness in the plates of the earth's crust.** *(Easy) (Skill 4.1)*

a. Faults.
b. Ridges.
c. Earthquakes.
d. Volcanoes.

Answer: a. Faults.

Faults are cracks in the earth's crust, that often cause earthquake results when the earth moves. Faults may lead to mismatched edges of ground, forming ridges, and ground shape may also be determined by volcanoes. The answer to this question must therefore be (A).

90.   **Which of the following is not a type of volcano?**
        (Average Rigor) (Skill 4.1)

a.   Shield volcanoes.
b.   Composite volcanoes.
c.   Stratus volcanoes.
d.   Cinder cone volcanoes.

Answer: c. Stratus Volcanoes.

There are three types of volcanoes. Shield volcanoes (A) are associated with non-violent eruptions and repeated lava flow over time. Composite volcanoes (B) are built from both lava flow and layers of ash and cinders. Cinder cone volcanoes (D) are associated with violent eruptions, such that lava is thrown into the air and becomes ash or cinder before falling and accumulating. Stratus' (C) is a type of cloud, not volcano, so it is the correct answer to this question.

91.   **Which of these is a true statement about loamy soil?**
        (Average Rigor) (Skill 4.1)

a.   Loamy soil is gritty and porous.
b.   Loamy soil is smooth and a good barrier to water.
c.   Loamy soil is hostile to microorganisms.
d.   Loamy soil is velvety and clumpy.

Answer: d. Loamy soil is velvety and clumpy.

The three classes of soil by texture are: Sandy (gritty and porous), Clay (smooth, greasy, and most impervious to water), and Loamy (velvety, clumpy, and able to hold water and let water flow through). In addition, loamy soils are often the most fertile soils. Therefore, the answer must be (D).

92.   **Lithification refers to the process that creates _____.**
       (Rigorous) (Skill 4.1)

a.   metamorphic rocks.
b.   sedimentary rocks.
c.   igneous rocks.
d.   lithium oxide.

Answer: b. Sedimentary rocks.

Lithification is the process of sediments coming together to form rocks, i.e. sedimentary rock formation. Metamorphic and igneous rocks are formed via other processes (heat and pressure or cooling magma or lava, respectively). Lithium oxide shares a word root with 'lithification' but is otherwise unrelated to this question. Therefore, the answer must be (B).

93.   **Which of the following is an example of an Igneous rock?**
       (Rigorous) (Skill 4.1)

a.   Quartz.
b.   Shale.
c.   Gneiss
d.   Obsidian

Answer: d. Obsidian

Quartz is a mineral. While it is often found in igneous rocks, it is not a rock. Shale is a sedimentary rock formed as fine sediment falls out of solution and piles up on the bottom of a lake or quiet body of water. Gneiss is a metamorphic rock formed from the addition of heat and pressure on an igneous rock such as granite. Obsidian is formed as lava cools very quickly as it is thrown out of a volcano. Therefore, the answer is (D).

94.   Fossils are usually found in _____ rock.
       (Easy) (Skill 4.2)

a.   igneous.
b.   sedimentary.
c.   metamorphic.
d.   larger grained.

Answer: b. Sedimentary

Fossils are formed by layers of dirt and sand settling around organisms, which is hardened over time, and taking an imprint of the organisms. When the organism decays, the hardened imprint is left behind. This is most likely to happen in rocks that form from layers of settling dirt and sand, i.e. sedimentary rock. Note that igneous rock is formed from molten rock from volcanoes (lava) or molten rock underground (magma) that cools slowly, while metamorphic rock can be formed from any rock under very high temperature and pressure changes. A description of the grain size of a rock is a way of classifiying the rock, but it does not tell the specific type of rock. The best answer is therefore (B).

95.   **The end of a geologic era is most often characterized by _____**
       (Average Rigor) (Skill 4.2)

a.   a general uplifting of the crust.
b.   the extinction of the dominant plants and animals
c.   the appearance of new life forms.
d.   all of the above.

Answer: d. All of the above.

Any of these things can be used to characterize the end of a geologic era, and often a combination of factors are applied to determining the end of an era.

96.   **Which of the following is the longest (largest) unit of geological time?**
      (Average Rigor) (Skill 4.2)

a.   Era
b.   Epoch.
c.   Period.
d.   Eon

Answer: d. Eon

Geological time is measured by many units, but the longest unit listed here (and indeed the longest used to describe the biological development of the planet) is the Eon. Eons are subdivided into Eras, which are subdivided into Periods, which are further divided into Epochs. Therefore, the answer is (D).

97.   **Which of the following is the best explanation of the fundamental concept of uniformitarianism?**
      (Rigorous) (Skill 4.2)

a.   The types and varieties of life will be seen in a see a uniform progression over time.
b.   The physical, chemical and biological laws that operated in the geologic past operate in the same way today.
c.   Debris from catastrophic events (i.e. volcanoes, and meteorites) will be evenly distributed over the effected area.
d.   The frequency and intensity of major geologic events will remain consistent over long periods of time.

Answer: b. The physical, chemical and biological laws that operated in the geologic past operate in the same way today.

While answers (A), (C), and (D) all could represent theories that have been proposed in geology, none of them accurately explain uniformitarianism. The general idea can be expressed, by the quote, "the present is the key to the past." The forces that we can observe today have been at work over most of Earth's history.

98. **The best preserved animal remains have been discovered in _____** *(Rigorous) (Skill 4.2)*

a.   resin
b.   shale
c.   tar pits
d.   glacial ice

Answer: c. tar pits.

Tar pits provide a wealth of information when it comes to fossils. Tar pits are oozing areas of natural asphalt, which were are sticky and often trap animals. These animals, without a way out, would die of starvation or be preyed upon. Their bones would remain in the tar pits, and be covered by the continued oozing of asphalt. Because the asphalt deposits are continuously added, the bones would not be exposed to much weathering. Some of the most complete and unchanged fossils from these areas include mammoths and saber toothed cats.

99. **The salinity of ocean water is closest to _____ .** *(Easy) (Skill 4.3)*

a.   0.035 %
b.   0.5 %
c.   3.5 %
d.   15 %

Answer: c. 3.5 %

Salinity, or concentration of dissolved salt, can be measured in mass ratio (i.e. mass of salt divided by mass of sea water). For Earth's oceans, the salinity is approximately 3.5 %, or 35 parts per thousand. Note that answers (A), (B) and (D) can be eliminated, because (A) and (B) are so dilute as to be hardly saline, while (D) is so concentrated that it would not support ocean life. Therefore, the answer is (C).

100. The theory of 'sea floor spreading' explains _____.
(Average Rigor) (Skill 4.3)

a. the shapes of the continents.
b. how continents collide.
c. how continents move apart.
d. how continents sink to become part of the ocean floor.

Answer: c. How continents move apart.

In the theory of 'sea floor spreading', the movement of the ocean floor causes continents to spread apart from one another. This occurs because crust plates split apart, and new material is added to the plate edges. This process pulls the continents apart, or may create new separations, and is believed to have caused the formation of the Atlantic Ocean. The answer is (C).

101. The theory of 'continental drift' is supported by which of the following?
(Average Rigor) (Skill 4.3)

a. The way the shapes of South America and Europe fit together.
b. The way the shapes of Europe and Asia fit together.
c. The way the shapes of South America and Africa fit together.
d. The way the shapes of North America and Antarctica fit together.

Answer: c. The way the shapes of South America and Africa fit together.

The theory of 'continental drift' states that many years ago, there was one land mass on the earth ('Pangaea'). This land mass broke apart via earth crust motion, and the continents drifted apart as separate pieces. This is supported by the shapes of South America and Africa, which seem to fit together like puzzle pieces if you look at a globe. Note that answer choices (A), (B), and (D) give either land masses that do not fit together, or those that are still attached to each other. Therefore, the answer must be (C).

102. Mount Kīlauea on the island of Hawaii, is a very active volcano that has continuous lava flow into the nearby ocean near it. What is the name of the type of shoreline created at the point where the lava flow meets the water?
(Rigorous) (Skill 4.3)

a. Stacking
b. Submerged
c. Developing
d. Emergent

Answer: d. Emergent

Answers (A) and (C) are not technical names for types of shorelines, although a stacked shoreline occurs when an island worn down to rocks. In this case the lava is building on previously deposited lava and although the lava itself is submerging under the water to develop the shoreline, the overall effect is the raising of the land out of the water. Thus the correct answer is (D).

103. Surface ocean currents are caused by which of the following?
(Rigorous) (Skill 4.3)

a. temperature.
b. changes in density of water.
c. wind.
d. tidal forces.

Answer: c. wind

A current is a large mass of continuously moving oceanic water. Surface ocean currents are mainly wind-driven and occur in all of the world's oceans (example: the Gulf Stream). This is in contrast to deep ocean currents which are driven by changes in density. Surface ocean currents are classified by temperature. Tidal forces cause changes in ocean levels, however they do not effect surface currents.

104. **Which of the following instruments measures wind speed?**
(Easy) (Skill 4.4)

a.   A barometer.
b.   An anemometer.
c.   A thermometer.
d.   A weather vane.

Answer: b. Anemometer.

An anemometer is a device to measure wind speed, while a barometer measures pressure, a thermometer measures temperature, and a weather vane indicates wind direction. This is consistent only with answer (B).

If you chose "barometer," here is an old physics joke to console you:

A physics teacher asks a student the following question:
        "Suppose you want to find out the height of a building, and the only tool you have is a barometer. How could you find out the height?"
        (The teacher hopes that the student will remember that pressure is inversely proportional to height, and will measure the pressure at the top of the building and then use the data to calculate the height of the building.)
        "Well," says the student, "I could tie a string to the barometer and lower it from the top of the building, and then measure the amount of string required."
        "You could," answers the teacher, "but try to think of a method that uses your physics knowledge from our class."
        "All right," replies the student, "I could drop the barometer from the roof and measure the time it takes to fall, and then use free-fall equations to calculate the height from which it fell."
        "Yes," says the teacher, "but what about using the barometer per se?"
        "Oh," answers the student, "I could find the building superintendent, and offer to exchange the barometer for a set of blueprints, and look up the height!"

105. **A closed contour line that has tiny comb-like lines along the inner edge indicates a _____**

(Average Rigor) (Skill 4.4)

a.   depression
b.   mountain
c.   valley
d.   river

Answer: a. depression

Contour lines are shown as closed circles in elevated areas and as lines with miniature perpendicular lined edges where depressions exist. These little lines are called hachure marks.

106. **In which layer of the atmosphere would you expect most weather conditions to occur?**
(Average Rigor) (Skill 4.4)

a.   Troposphere
b.   Thermosphere
c.   Mesosphere
d.   Stratosphere

Answer: a. Troposphere

The troposphere is the lowest portion of the Earth's atmosphere. It contains the highest amount of water and aerosol. Because it touches the Earth's surface features, friction builds. For all of these reasons, weather is most likely to occur in the troposphere.

107. **The transfer of heat from the earth's surface to the atmosphere is called**
   (Rigorous) (Skill 4.4)

a. convection.
b. radiation.
c. conduction.
d. advection.

Answer: c. conduction

Radiation is the process of warming through rays or waves of energy, such as the Sun warms earth. The Earth returns heat to the atmosphere through conduction. This is the transfer of heat through matter, such that areas of greater heat move to areas of less heat in an attempt to balance temperature.

108. **Which of the following causes the aurora borealis?**
   (Rigorous) (Skill 4.4)

a. Particles from the sun
b. Gases escaping from earth
c. Particles from the moon
d. Electromagnetic discharges from the North pole.

Answer: a. particles from the sun

Aurora Borealis is a phenomenon caused by particles escaping from the sun. The particles escaping from the sun include a mixture of gases, electrons and protons, and are sent out at a force that scientists call solar wind. Together, we have the Earth's magnetosphere and the solar wind squeezing the magnetosphere and charged particles everywhere in the field. When conditions are right, the build-up of pressure from the solar wind creates an electric voltage that pushes electrons into the ionosphere. Here they collide with gas atoms, causing them to release both light and more electrons.

109. **Which of the following units is not a measure of distance?**
     (Easy) (Skill 4.5)

a.   AU (astronomical unit).
b.   Light year.
c.   Parsec.
d.   Lunar year.

Answer: d. Lunar year.

Although the terminology is sometimes confusing, it is important to remember that a 'light year' (B) refers to the distance that light can travel in a year. Astronomical Units (AU) (A) also measure distance, and one AU is the distance between the sun and the earth. Parsecs (C) also measure distance, and are used in astronomical measurement- they are very large, and are usually used to measure interstellar distances. A lunar year, or any other kind of year for a planet or moon, is the time measure of that body's orbit. Therefore, the answer to this question is (D).

110. **The phases of the moon are the result of its _____ in relation to the sun.**
     (Average Rigor) (Skill 4.5)

a.   revolution
b.   rotation
c.   position
d.   inclination

Answer: c. position

The moon is visible in varying amounts during its orbit around the earth. One half of the moon's surface is always illuminated by the Sun (appears bright), but the amount observed can vary from showing a full moon to no illumination at all.

111. **Which of the following is a correct explanation for an astronaut's 'weightlessness' while in orbit?**
(Average Rigor) (Skill 4.5)

a. Astronauts continue to feel the pull of gravity in space, but they are so far from the planets that the force is small.

b. Astronauts continue to feel the pull of gravity in space, but spacecrafts have such powerful engines that those forces dominate, reducing effective weight.

c. Astronauts do not feel the pull of gravity in space, because space is a vacuum.

d. The cumulative gravitational forces, that the astronaut is experiences from all sources in the solar system equal out to a net gravitational force of zero.

Answer: a. Astronauts continue to feel the pull of gravity in space, but they are so far from planets that the force is small.

Gravity acts over tremendous distances in space (theoretically, infinite distance, though certainly at least as far as any astronaut has traveled). However, gravitational force is inversely proportional to distance squared from a massive body. This means that when an astronaut is in space, s/he is far enough from the center of mass of any planet that the gravitational force is very small, and s/he feels 'weightless'. Space is mostly empty (i.e. vacuum), and spacecrafts have powerful engines. However, none of these has the effect attributed to it in the incorrect answer choices (B), or (C). Although, theoretically there is a point in space where the cumulative gravitational forces of sources within the solar system would equal a net force of zero, that point would be in constant motion and difficult to find, making answer D unlikely at best and but more accurately near impossible to keep an astronaught at this point. The answer to this question must therefore be (A).

112. **The planet with retrograde rotation is**
(Rigorous) (Skill 4.5)

a. Pluto
b. Uranus
c. Venus
d. Saturn

Answer: c. Venus

Venus has an axial tilt of only 3 degrees and a very slow rotation. It spins in the direction opposite of its counterparts (who spin in the same direction as the Sun). Uranus is also tilted and orbits on its side. However, this is thought to be the consequence of an impact that left the previously prograde rotating planet tilted in such a manner.

113. **What is the main difference between the 'condensation hypothesis' and the 'tidal hypothesis' for the origin of the solar system?** *(Rigorous) (Skill 4.5)*

a. The tidal hypothesis can be tested, but the condensation hypothesis cannot.
b. The tidal hypothesis proposes a near collision of two stars pulling on each other, but the condensation hypothesis proposes condensation of rotating clouds of dust and gas.
c. The tidal hypothesis explains how tides began on planets such as Earth, but the condensation hypothesis explains how water vapor became liquid on Earth.
d. The tidal hypothesis is based on Aristotelian physics, but the condensation hypothesis is based on Newtonian mechanics.

Answer: b. The tidal hypothesis proposes a near collision of two stars pulling on each other, but the condensation hypothesis proposes condensation of rotating clouds of dust and gas.

Most scientists believe the 'condensation hypothesis,' i.e. that the solar system began when rotating clouds of dust and gas condensed into the sun and planets. A minority opinion is the 'tidal hypothesis,' i.e. that the sun almost collided with a large star. The large star's gravitational field would have then pulled gases out of the sun; these gases are thought to have begun to orbit the sun and condense into planets. Because both of these hypotheses deal with ancient, unrepeatable events, neither can be tested, eliminating answer (A). Note that both 'tidal' and 'condensation' have additional meanings in physics, but those are not relevant here, eliminating answer (C). Both hypotheses are based on best guesses using modern physics, eliminating answer (D). Therefore, the answer is (B).

114. **Which of the following is a true statement about radiation exposure and air travel?**
(Average Rigor) (Skill 5.1)

a. Air travel exposes humans to radiation, but the level is not significant for most people.
b. Air travel exposes humans to so much radiation that it is recommended as a method of cancer treatment.
c. Air travel does not expose humans to radiation.
d. Air travel may or may not expose humans to radiation, but it has not yet been determined.

Answer: a. Air travel exposes humans to radiation, but the level is not significant for most people.

Humans are exposed to background radiation from the ground and in the atmosphere, but these levels are not considered hazardous under most circumstances, and these levels have been studied extensively. Air travel does create more exposure to atmospheric radiation, though this is much less than people usually experience through dental X-rays or other medical treatment. People whose jobs or lifestyles include a great deal of air flight may be at increased risk for certain cancers from excessive radiation exposure. Therefore, the answer is (A).

115. **Genetic engineering has benefited agriculture in many ways. Which of the following is not one of these benefits?**
(Rigorous) (Skill 5.1)

a. Developing a bovine growth hormone to increase milk production.
b. Strains of crops have been developed to resist herbicides.
c. The development of micro-orgasms that breakdown toxic substances into harmless compounds.
d. Genetically vaccinating plants against viral attack.

Answer: c. The development of micro-orgasms to breakdown toxic substances into harmless compounds.

All of the answers are actual results of genetic engineering, however only answer (C) has not been used for agricultural purposes. These micro-organisms have however been used at the sites of oil spills and toxic waste sites.

116. **Which of the following is not a common type of acid found in 'acid rain' or acidified surface water?**
(Average Rigor) (Skill 5.2)

a.   Nitric acid.
b.   Sulfuric acid.
c.   Carbonic acid.
d.   Hydrofluoric acid.

Answer: d. Hydrofluoric acid.

Acid rain forms predominantly from pollutant oxides in the air (usually nitrogen-based NOx or sulfur-based SOx), which become hydrated into their acids (nitric or sulfuric acid). Because of increased levels of carbon dioxide pollution, carbonic acid is also commonly found in acidified surface water environments. Hydrofluoric acid is sometimes present, but it is much less common. In general, carbon, nitrogen, and sulfur are much more prevalent in the environment than fluorine. Therefore, the answer is (D).

117. **Which of the following activities is least likely to lead to a disasterous event?**
(Rigorous) (Skill 5.2)

a.   The use of vehicles that emit greenhouse gases.
b.   Housing development on potentially unstable ground.
c.   The use of nuclear power plants.
d.   Strip mining.

Answer: c. The use of nuclear power plants.

Although many people are afraid of having nuclear power plants near their homes, in reality nuclear power has a good track record for several decades, with minimal impact on the environment. Answers (A) and (D) slowly alter the environment and can cause a variety of harmful effects. Answer (B) is an example of the times in California, where homes have been swallowed by sink holes, or slid down the sides of hills. Thus answer (C) is the best answer.

118. **Contamination may enter groundwater by which of the following means?**
(Easy) (Skill 5.3)

a. air pollution
b. leaking septic tanks
c. photochemical processes
d. sewage treatment plants.

Answer: b. leaking septic tanks

Leaking septic tanks allow contamination to slowly seep into the ground, where it is absorbed into the water table and infects the groundwater. The only other reasonable possibility is sewage treatment plants, which isolate the waste from ground water until it has reached a state that it will not be hazardous to release into ground or surface water.

119. **Which of the following is the most accurate definition of a non-renewable resource?**
(Average Rigor) (Skill 5.3)

a. A non-renewable resource is never replaced once used.
b. A non-renewable resource is replaced on a timescale that is very long relative to human life-spans.
c. A non-renewable resource is a resource that can only be manufactured by humans.
d. A non-renewable resource is a species that has already become extinct.

Answer: b. A nonrenewable resource is replaced on a timescale that is very long relative to human life-spans.

Renewable resources are those that are renewed, or replaced, in time. Examples include fast growing plants, animals, or oxygen gas. (Note that while sunlight is often considered a renewable resource, it is actually a non-renewable but extremely abundant resource.) Non-renewable resources are those that renew themselves only on very long timescales, usually geologic timescales. Examples include minerals, metals, or fossil fuels. Therefore, the correct answer is (B).

120. **All of the following are potential sources of alternative energy that are currently being used except**
    (Rigorous) (Skill 5.3)

a. Biomass energy
b. Geothermal energy
c. Wind energy
d. Nuclear fusion power plants.

Answer: d. Nuclear Fusion power plants.

Biomass power generation is being used at many landfill sites. The country of Iceland uses geothermal energy extensively, and the country of Denmark hopes to have an off shore wind turbine farm supplying most if not all of its power in the next few years. Although the hope would be to one day be able to use fusion power plants, to this date no one has been able to maintain a stable fusion reaction. Currently, all nuclear power plants are fission based. So the answer is (D).

121. **Which of the following is not considered ethical behavior for a scientist?**
    (Easy) (Skill 5.4)

a. Citing the sources before data is published.
b. Publishing data before other scientists have had a chance to replicate results.
c. Collaborating with other scientists from different laboratories.
d. Publishing work with an incomplete list of citations.

Answer: d. Publishing work with an incomplete list of citations.

One of the most important ethical principles for scientists is to cite all sources of information, data and analysis when publishing work. It is reasonable to use unpublished data (A), as long as the source is cited. Most science is published before other scientists replicate it (B), and frequently scientists in the same or different laboratories (C). These are all ethical choices. However, publishing work without the appropriate citations, is unethical. Therefore, the answer is (D).

122. **Which of the following is the least ethical choice for a school laboratory activity?**
(Average Rigor) (Skill 5.4)

a. A genetics experiment tracking the fur color of mice.
b. Dissection of a preserved fetal pig.
c. Measurement of goldfish respiration rates at different temperatures.
d. Pithing a frog to observe the circulatory system.

Answer: d. Pithing a frog to observe the circulatory system.

Scientific and societal ethics make choosing experiments in today's science classroom difficult. It is possible to ethically perform choices (A), (B), or (C), if due care is taken. (Note that students will need significant assistance and maturity to perform these experiments). However, modern practice precludes pithing animals (causing partial brain death while allowing some systems to function) as inhumane. Therefore, the answer to this question is (D).

123. **Which of the following lines of scientific research has not been strictly regulated by the federal government?**
(Rigorous) (Skill 5.4)

a. Stem-cell research
b. Nuclear physics
c. Animal testing
d. Microcomputing development

Answer: d. Microcomputing development

Answers (A) and (C) have seen some time in the public spotlight in recent years and have been singled out for regulations as to what is appropriate (i.e. ethical) research. Answer (B) has been regulated almost since the creation of the field, primarily for safety concerns. Microcomputing development has seen little regulation by the government in how they conduct their research, however patent and FCC rules do come into effect when microcomputing research leads to products, so answer (D) is the best answer.

124. **Extensive use of antibacterial soap has been found to increase the virulence of certain viral strains infections in hospitals. Which of the following might be an explanation for this phenomenon?**
   (Average Rigor) (Skill 5.5)

   a.   Antibacterial soaps do not kill viruses.
   b.   Antibacterial soaps do not incorporate the same antibiotics used as medicine.
   c.   Antibacterial soaps kill a lot of bacteria, and only the hardiest ones survive to reproduce.
   d.   Antibacterial soaps can be very drying to the skin.

Answer: c. Antibacterial soaps kill a lot of bacteria, and only the hardiest ones survive to reproduce.

All of the answer choices in this question are true statements, but the question specifically asks for a cause of increased disease virulence in hospitals. This phenomenon is due to natural selection. The bacteria that can survive contact with antibacterial soap are the strongest ones, and without other bacteria competing for resources, they have more opportunity to flourish. This problem has led to several antibiotic-resistant bacterial diseases in hospitals nation-wide. Therefore, the answer is (C). However, note that answers (A) and (D) may be additional problems with over-reliance on antibacterial products.

125. **Which is not an example of how the general public is learning more about health issues?**
   (Rigorous) (Skill 5.5)

   a.   Childhood Vaccinations
   b.   Nutrition Education
   c.   Radical diet plans
   d.   Education about signs and symptoms of heart attacks.

Answer: c. Radical diet plans

While all of the answers are prevalent in public notice, information about childhood vaccinations and education about the signs and symptoms of heart attacks are well-received and understood. Unlike nutrition education, radical diet plans entice the public with promises of quick fixes for weight problems and rarely deliver. Thus, radical diet plans are not teaching the public anything valuable about health issues.

XAMonline, INC. 21 Orient Ave. Melrose, MA 02176

Toll Free number 800-509-4128

TO ORDER Fax 781-662-9268 OR www.XAMonline.com

WEST SERIES

PO#   Store/School:

Address 1:

Address 2 (Ship to other):

City, State Zip

Credit card number_____-_____-_____-_____   expiration_____

EMAIL _____

PHONE   **FAX**

ISBN	TITLE	Qty	Retail	Total
978-1-58197-638-0	WEST-B Basic Skills			
978-1-58197-609-0	WEST-E Biology 0235			
978-1-58197-565-9	WEST-E Chemistry 0245			
978-1-58197-566-6	WEST-E Designated World Language: French Sample Test 0173			
978-1-58197-557-4	WEST-E Designated World Language: Spanish 0191			
978-1-58197-614-4	WEST-E Elementary Education 0014			
978-1-58197-636-6	WEST-E English Language Arts 0041			
978-1-58197-634-2	WEST-E General Science 0435			
978-1-58197-637-3	WEST-E Health & Fitness 0856			
978-1-58197-635-9	WEST-E Library Media 0310			
978-1-58197-674-8	WEST-E Mathematics 0061			
978-1-58197-556-7	WEST-E Middle Level Humanities 0049, 0089			
978-1-58197-568-0	WEST-E Physics 0265			
978-1-58197-563-5	WEST-E Reading/Literacy 0300			
978-1-58197-552-9	WEST-E Social Studies 0081			
978-1-58197-639-7	WEST-E Special Education 0353			
978-1-58197-633-5	WEST-E Visual Arts Sample Test 0133			
	SUBTOTAL		Ship	$8.25
	FOR PRODUCT PRICES VISIT WWW.XAMONLINE.COM		TOTAL	

e, TN USA
2010

V00002B/28/P